VOLUME I: GEOLOGIC HISTORY OF ILLINOIS

Steven D.J. Baumann, P.G., Teresa Arrospide, and Jamie L. Bardwell

Edition 2: ©2020

Midwest Institute of Geosciences and Engineering

(inside cover)

I0515710

We would like to personally thank the following individuals who have made this book possible by either joining us in the field, selecting the stops, editing, or just by encouraging my love for rocks.

I dedicate this book to all of you!

David H. Malone

Micaela Krol

Sarah M. Hall

Alexandra B. Cory

David Johnson

Mary M. Pryjda

Tiffany E. Mikes

Cover photo taken by Steven Baumann. Maps adapted from the Illinois State Geologic Survey's online maps. "Bedrock Geologic Map of Illinois" (2005) and "Quaternary (Ice Age) Deposits" (2005).

VOLUME I
GEOLOGIC HISTORY OF ILLINOIS

STEVEN D.J. BAUMANN, P.G.
TERESA ARROSPIDE
JAMIE L. BARDWELL

Midwest Institute of Geosciences and Engineering

© 2017 and 2020

www.mige-web.org

EDITED BY: **Sandra K. Dylka**

FIGURES AND DIAGRAMS: **Teresa Arrospide**

PHOTOS TAKEN BY: **Steven D.J. Baumann**

Table of Contents

Preface for Volumes I and II..........1

Introduction 1

Volume I: Geologic History of Illinois 4

Evolution of the Earth through Time..........5

    -Plate Tectonics and the Rock Cycle..........5

    -Sedimentary Rocks and Systems..........7

    -Geological Structures..........9

    -Geological Time and Formations..........14

    -Geomorphology and Landforms..........16

Introduction to the Physical History of the Prairie State..........20

    -Mesoproterozoic and Neoproterozoic Eras: Illinois' Roots and the Great Unconformity..........20

    - Paleozoic Era: Appearance of Vast Marine Life..........22

        - Cambrian Period: Explosion of Life..........22

        - Ordovician Period: Time of Crinoids, Brachiopods, and Glaciers..........24

        - Silurian and Devonian Periods: Great Reefs and Clear Seas..........27

        - Mississippian, Pennsylvanian, and Permian Periods: Time of Shallow Marine Pools, Coal Swamps, and Volcanic Intrusions 30

-Mesozoic Era: Land Animals Reach Their Apex ……….34

       -Triassic and Jurassic Period: The Rocks that Never Were ……….34

       - Cretaceous Period: Illinois was a State with Ocean Front Property ……….35

- Cenozoic Era: A New Dynasty Dawns ……….36

       -Paleogene and Neogene: Illinois Begins to Take on a Familiar Appearance ……….36

       -Quaternary: Age of Ice and Men ……….38

- Geobits (8) ……….45

-Glossary………52

-Additional Reading……….59

-References……….60

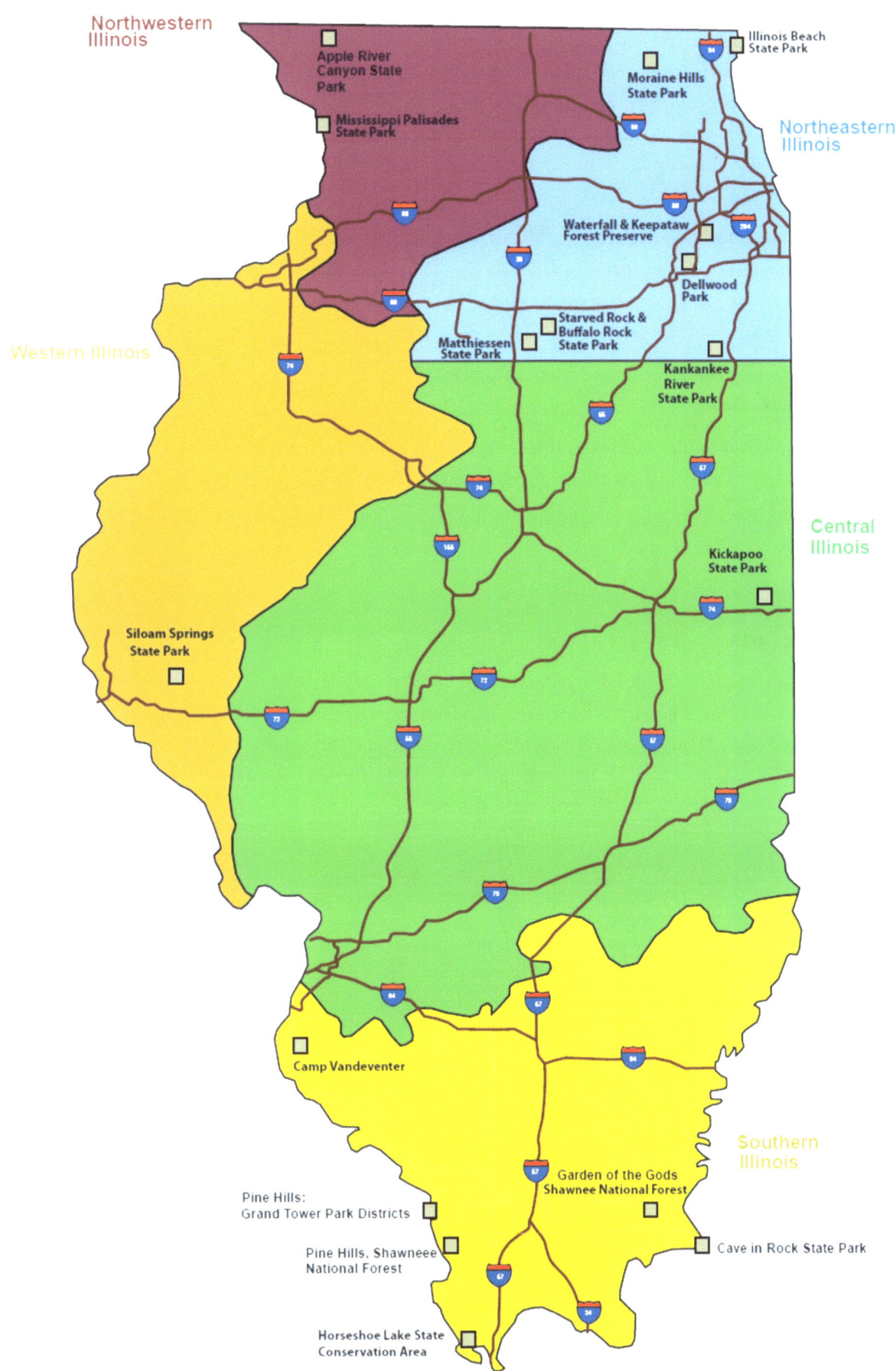

BOOK DIVISIONS: This map depicts the areas divided out in this book. Also shown are the parks included in Volume II.

Preface for Volumes I and II

This book was written for the any person interested in the geology of Illinois, regardless of background. We attempted to keep the technical lingo to a minimum. Geology, like all sciences has its own set of vocabulary. If technical words were unavoidable, we attempt to explain them in either the text or the glossary. It is the first of two volumes. In this volume we explore the background of how Illinois came to be. In the second volume we will take you to specific locations to observe the rocks in outcrop.

Volumes I and II are the second edition of a single book that was called "Roadside and Riverside Geology of Illinois". We wanted to change some things. The original 2017 edition was too long. It contained over 170 color photos and about 30 color diagrams and figures. We wanted to add a couple more. That increased the cost on an already expensive book. So the decision was made to break it into two volumes.

In this volume we take you through the geologic history of Illinois. From its fiery Precambrian roots to the modern landscape carved by ice and water.

The best times of the year to look at the geology of Illinois is late winter to late spring after the snow has melted and before the trees grow their leaves. Fall and early winter are also excellent times to view the geology.

We hope you find this book informative and enlightening. Illinois is a remarkable state and is so much more than a land of mega farms and cities.

Introduction

Perspective on How the Earth Has Changed Through Time

Currently, the Illinois landscape is made of flat to rolling farms, deep river valleys, home to one of the largest cities on the planet, and a freshwater lake so large it spans the horizon. Illinois is known as the "Prairie State", although there aren't many prairies around these days. It was originally settled by native people coming down from the northwest and probably from the east more than 12,000 years ago. The Europeans settled in Illinois much later, and in 1818 Illinois became the 21st state. Long before any of this happened, Illinois was very different.

When we look outside at the surrounding Illinois' countryside, there is a tendency to think that the landscape we see today has always been that way. As human beings, with our limited life spans, we just take this assumption as fact. The truth couldn't be further from our perception. The Earth is estimated to be at least 4.54 billion years old. This amount of time is difficult to comprehend. We can think of time as a distance and time. Imagine an 80 year old person. If one inch on the ground was equal to one year, an 80 year old person's life would be 6 feet and 8 inches from our starting point. The timespan of the human race is longer but would still only be about one and a half miles from your starting point. In comparison, the age of the Earth would be about 71,650 miles from your starting point. This is equivalent to just slightly less than three full trips around the Earth at the equator!

Our perception of the world around us is very limited and confining at best. We are restricted to what we experience during our short stay on this borrowed planet. This narrow experience can lead to many misconceptions about the Earth. It can also lead to other more noticeable and deadly impacts. We tend to view resources that are presently abundant as always being abundant. If we continue to perceive our resources in this way, we are sure to exhaust them. Perhaps the two most endangered resources are oil and water. Mostly due to overuse, mismanagement, and the ingrained perception that these two resources will always be economically viable. Such a limited view of our planet can lead to engineering mistakes as well. The devastating effects of the New Madrid earthquakes of 1811 and 1812 should have taught us something, it's that we should reinforce tall structures in the earthquake prone areas of the mid-continent. Yet very few structures in the greater Saint Louis area could withstand another series of earthquakes like the ones in 1811 and 1812. Our view of the Earth upon which we live from the perspective of our lifespan, is akin to trying to study the stars by looking at a single patch of sky through a cardboard tube. You can get a sense of what's out there but you're missing the big picture.

The study of geology teaches us that the Earth is always changing. It may not be changing at the pace a person can observe, but it is always dynamic and in motion. A lot can happen in 4.54 billion years! The study of geology and other earth sciences is more than purely academic. The study of the Earth breaks us out of our limited view of the planet that we inhabit. We can figure out the story of our planet by looking at the geology. After all, the history of the Earth isn't in a museum, it's in the rocks.

Earth may be old but Illinois isn't as old as the planet, even at its deep Precambrian roots. The oldest rocks, which are deep under the surface of Illinois, only go back about 1.6 billion years. 90% of the geology at the surface in Illinois is less than one million years old. The surface of Illinois is young compared to the age of our planet. Don't let the calm appearance of the landscape fool you into thinking Illinois is a bland and unchanging place. If we go back as little as 15,000 years the landscape was much more barren and cold, full of lakes that no longer exist, and large mammals now extinct. Lake Michigan, as we know it, did not exist.

We don't need to go back that far in order to see that the landscape of Illinois can change. In the narrow timespan of a couple of centuries, the surface can change. During the New Madrid earthquakes in the early 19th century, the course of the Mississippi River was altered in a single event. In 1993 the Mississippi River almost added the area south of Miller City (in Alexandria County) to Missouri, when it attempted to alter its course and abandon its active meander. If this cutoff would have been successful, about 20 square miles of Illinois would now lie south of the Mississippi. The Ohio River in Southern Illinois altered its course in places after the flooding in 2011. Even on a small and recent scale, the Prairie State isn't quiet. Although water is the main driving force for shaping Illinois, that wasn't always the case.

The further we go back, the more things change and become unfamiliar, and you don't have to travel very far. If you go back even 20,000 years ago, vast ice sheets, hundreds of feet thick, covered nearly 1/2 of Illinois. Trees within 30 miles of the ice would have been rare. The landscape probably looked more like the southwest coast of Greenland today. Just not with as much topographic relief or the deep fjords cut into the sea. Illinois was actually further from the ocean 20,000 years ago then it is today, due to a global lowering of sea level caused by the most recent ice advance (the Wisconsin Episode).

Travel back about three million years ago and Illinois was a land of vast forests and high hills with scattered grasslands. The landscape was more comparable to the modern landscape of Tennessee or the Galena area in northwest Illinois. The major tributary emptying into the Mississippi River from the east was not the Ohio River, as it is today. It was a river valley with headwaters in West Virginia and entering Illinois in Iroquois and Vermilion Counties.

The river occupied the Mahomet-Teays (or just Mahomet) Bedrock Valley. This now buried river valley rivaled the modern Mississippi River in size. The Mississippi River itself did not follow its present course. During these pre-ice age times, it flowed southeast through Rock Island County, linking up with the modern, north-south portion of the Illinois River in Putnam County.

If we go back 55 million years, Illinois was probably only slightly hillier than it is today. What we presently call the Gulf of Mexico indurated the southern tip of Illinois. At this time the planet had no ice caps. It was a landscape of forests and primitive mammals, including the ancestor of the modern horse.

75 million years ago, Illinois was flatter than it is today. The dinosaurs walked the state amidst scattered forests and savannas with two oceans nearby. During this time, a vast north-south trending sea called the Western Interior Seaway covered many of our neighboring states like Iowa and Missouri to the west. It also covered most of the states that make up the Great Plains, north into Canada, and reaching the eastern half of Alaska. The other great sea extended from the south coming as far north as Jefferson County and was connected to the Western Interior Seaway southwest of the Ozarks in Missouri. This southern ocean covered all of Florida, Mississippi, Louisiana, Alabama, and most of Tennessee and Georgia. The Mississippi River did not exist at this time. The major rivers into these now dead seaways probably came from the northeast and east. Illinois was located a little further south than it is today. Downtown Chicago sat at about 35° north latitude. Today it is located at about 41.9° north.

If we go back 300 million years, Illinois was on the equator. The topography was close to sea level and covered with vast swamps that were fed by broad meandering rivers coming from the north and northeast. During this time a unique creature, occupied the swamps. It is affectionately known as the "Tully Monster" and is unique to one geologic formation in Illinois.

If people were around 420 million years ago, Illinois might have been a summer tourist destination. The entire state was covered by shallow tropical seas with reefs that would rival the modern Great Barrier Reefs off the northeastern coast of Australia. There would have only been small islands and atolls peaking above the quiet waves. Today these reefs are quarried as a source of stone used as track ballast and road base throughout the Midwest.

600 million years ago, long before animals or plants covered the land, Illinois was a barren and lifeless landscape of granitic hills, similar in size and shape to the modern Ozark Mountains of Missouri. The rocks that made up the hills are what we presently refer to as the "Eastern Granite-Rhyolite Province" These ancient hills are not presently exposed at the surface anywhere in Illinois. The closest they come to the surface is about 1,500 feet in North-central Illinois and are referred to as basement rocks.

Any further back in time and we can only speculate what Illinois looked like. There is a gap in the rock record recognized throughout North America as the "Great Unconformity". It is a missing span of time in the rock record. This span of missing time, or "unconformity", is anywhere from 100 million years to 1.1 billion years, depending on where you are on the continent. It is possible that the area of the land presently occupied by Illinois, did not exist at all prior to the emplacement of the basement rock 1.6 billion years ago.

VOLUME I
Geologic History of Illinois

Here we explore the forces that shaped not only Illinois, but the Earth as a whole. We will take you on the driving forces that shape Earth trough time, as well as the geologic time spans that have shaped Illinois.

Evolution of Earth Through Time

Plate Tectonics and the Rock Cycle

Earth is a very dynamic and always changing planet. Most of what the Earth does is so slow and quiet, that we don't even notice. Most processes that shape the Earth take thousands of generations to notice. What drives the forces of the Earth? How did mountains, oceans, rivers, and continents form? The short answer…Plate Tectonics.

Plate Tectonics is the side effect of how our planet loses its internal heat. Plate Tectonics comes from a concept that was called "continental drift" when it was first proposed over 100 years ago. But continental drift is not Plate Tectonics. We did not begin to understand how the continents moved until the 1960's. As the Earth loses its internal heat, which is left over from its formation 4.54 billion years ago, and radioactive decay. Magma plumes rise to the surface and force the land apart. This process of splitting the continents causes them to move away from one another as new ocean crust forms between them at the spreading zones. Since the Earth is not getting any larger, something has to happen to the old ocean crust. The older ocean crust becomes heavy and breaks. This causes it to sink back into the Earth's interior in subduction zones. The old ocean crust eventually gets recycled into the mantle, but the continents do not. Over time, Plate Tectonics causes continents to split apart, eventually come back together, then split apart again, and so on. The continents are lighter than the ocean crust. As a result, they do not get recycled down into the mantle. They move about the surface, along for the ride. When the continents come together in a large single or a couple of continents, we call them supercontinents. Most people are familiar with the most recent supercontinent of Pangaea, but Earth has had at least one to five other such massive continents, going back three billion years. These continents assemble, break apart, and reassemble over millions of years. Plate Tectonics and the shifting of Earth's crust is the cause of tsunamis, earthquakes, and volcanic eruptions.

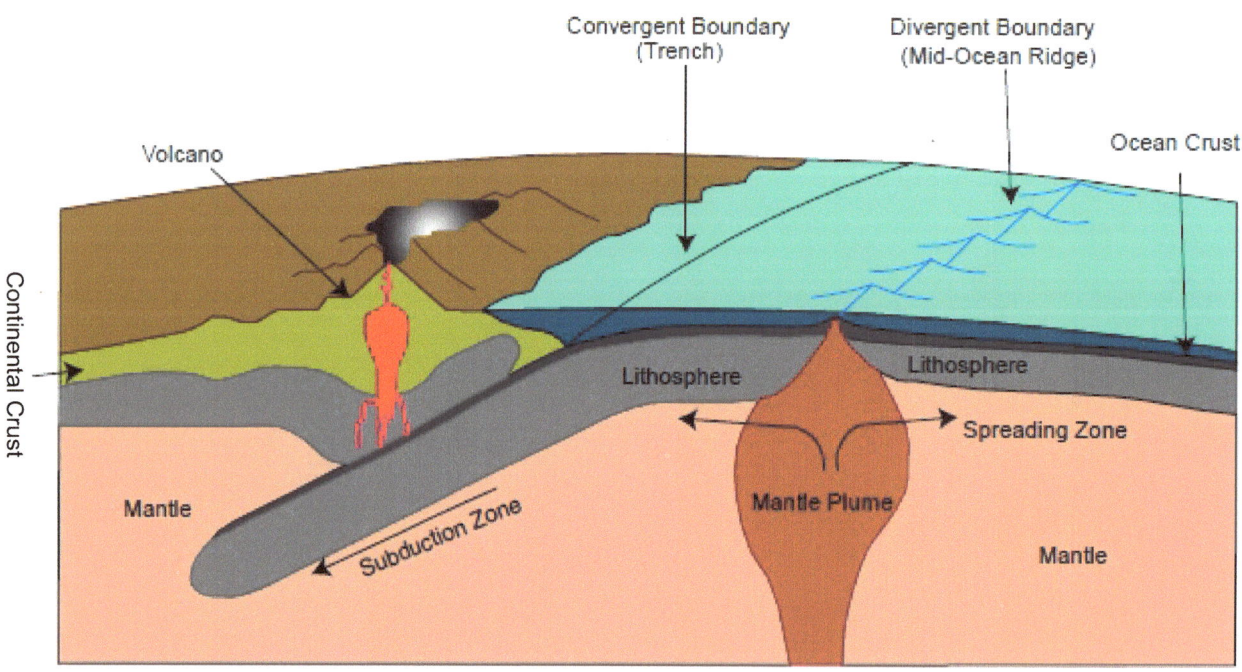

PLATE TECTONICS: *The driving forces of geology are seeded deep inside the Earth. Magma wells upwards causing plates to move away from one another. In subduction areas, the ocean plates are brought back down into the planet's interior. The continents do not get subducted. They are just move along the surface. When continents meet in an area where the ocean is subducted large mountain ranges form in an event called an orogeny, and eventually the process of subduction stops and resumes elsewhere.*

Earth is the only known planet in the solar system to have Plate Tectonics but, this has not always been the case. Before about 2.5 to 3.2 billion years ago something other than Plate Tectonics was going on. We know this because of the rocks from this time. Before this time we didn't have true "granites". Coarse igneous felsic rocks from this time tend to be very rich in plagioclase and deficient in alkali-feldspars or K-spar. What dominated were mostly tonalites and diorites. Sedimentary continental shelf deposits are lacking before 2.5 billion years ago. From the few sedimentary rocks older than 2.5 billion years, we know that the Earth has had oceans for more than four billion years and single celled life appears in the fossil record about 3.9 billion years ago. As for how the Earth lost its heat before 2.5 billion years ago, we don't know. The planet Venus may hold the answer. Venus is similar in size and mass to the Earth. Yet, it is very unearth like. Venus appears to lose its internal heat by massive igneous resurfacing events that may occur every 300 million to 600 million years. Since Venus has no oceans or a moon to interact tidally with, this "pre-plate tectonic" mechanism is still occurring. Earth may have had a similar way of losing its heat early in its history, when the crust and lithosphere were thinner.

Plate tectonics is the mechanism for what drives what geologists call the "Rock Cycle". All rocks on Earth are either crystalized magma or lava from the Earth's interior (igneous), rocks formed deep in the crust, from heat and compression (metamorphic), or recycled rocks that have been deposited chemically, organically, or by wind, water, or ice (sedimentary). Igneous, metamorphic, and sedimentary rocks are the three types of rock that occur on Earth. These rocks all change over time. Their formation is driven by the internal workings of the planet (Plate Tectonics), by surface mechanisms (erosion), or extraterrestrial events such as meteor impacts and tidal forces. Other than the small Permian dikes in Sothern Illinois, all rocks at the surface in Illinois are sedimentary. At the surface, erosion is the dominant force that shapes rocks. Erosion doesn't only wash rock away; it also deposits it elsewhere to become new rock.

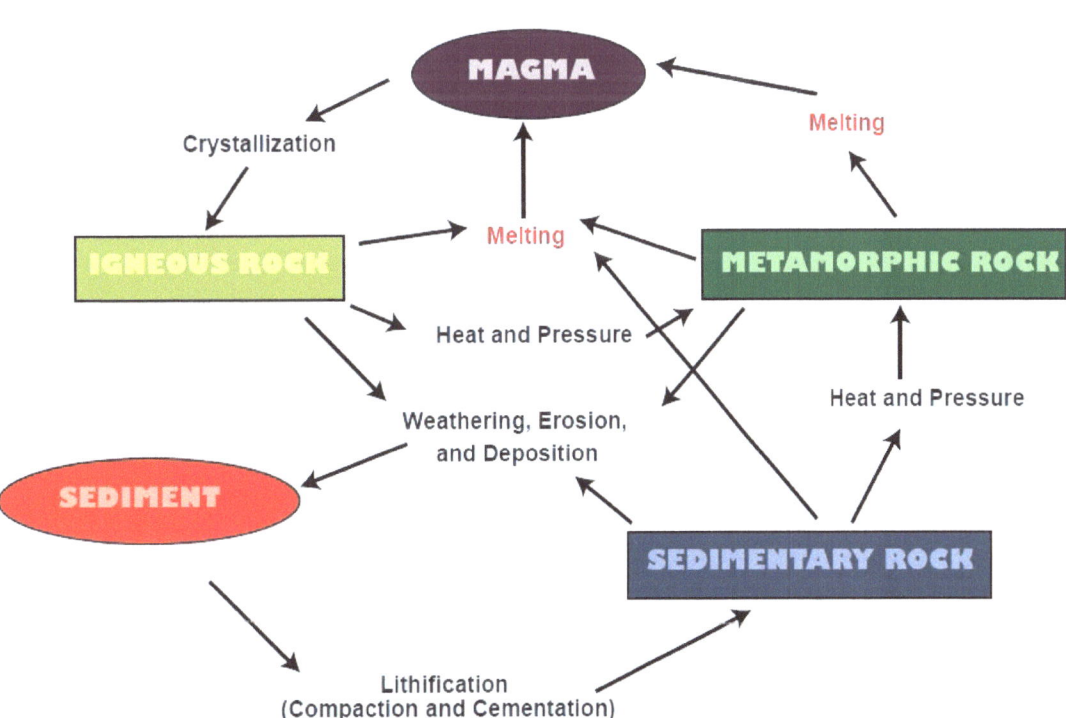

THE ROCK CYCLE: *The rock cycle is a key concept to understanding the evolution and identification of the three major types of rocks. The three major types of rocks are in the square boxes. If you follow the arrows, they show how one type of rock can become another type.*

There are no metamorphic or igneous (minus the dikes in Southern Illinois) rocks present at the surface in Illinois except for the ones carried here by the glaciers of the last Ice Age from Minnesota, Michigan, Wisconsin, and Canada as boulders and cobbles of glacial erratics within the drift and till. See the chart on page 9 for how the different types of rock are distributed in Illinois.

Sedimentary Rocks and Systems

There are three main types of sedimentary rocks and all are present in Illinois. Clastic sedimentary rocks (i.e. sandstone, shale, diamicton, siltstone, and conglomerate) are formed from rocks that have been eroded by either water, ice, or wind, and redeposited. Chemical rocks (i.e. limestone, dolostone, ironstones, and anhydrites) are deposited directly from water (or the creatures that live within the water). Organic rocks (i.e. coal and certain limestones and dolostones) are deposited when plants and animals die, and build up layers of peat and carbonate rich layers or in beds of dead skeletal remains as in reefs. Over time the peat can become buried by younger sediments and gets turned into coal, or the reef becomes buried and hardens. This is called lithification.

Clastic rocks are formed by the mechanical and chemical weathering of other rocks. The most common type of clastic rock in Illinois is shale. Shale is the rock version of clay. It is formed when water alters minerals in other rocks and rearranges the molecules. The most common type of clay is called illite, which is named for Illinois. Shales are abundant in Pennsylvanian and Ordovician rocks.

Sandstone makes up another abundant clastic rock type in Illinois. It forms from the mechanical weathering of other rocks. Sandstone contains sand sized particles of sedimentary origin. The sizes of the grains tell you nothing of the composition of sandstone. Sandstone is usually made of quartz in older rocks in Illinois, such as the Saint Peter and New Richmond Formations, but not all sandstone is quartz dominated. There are feldspar rich sands (also called arkoses), which are rare in Illinois. Feldspar rich sands are common in places near igneous rocks serve as a source for the sand. All of the igneous rocks that could be the source of feldspar rich sandstones were buried half a billion years ago. Most of the glacial sands in Illinois have large amounts of lithic grains. Lithic sands are made of a multitude of other rocks, but the grains are sand sized. The most common lithic sands in Illinois contain igneous fragments as well as chert, shale, and carbonate grains.

The most common chemical rocks in Illinois are carbonate rocks which are represented by limestone and dolostone in Illinois. Carbonate rocks are usually deposited straight from sea water, the decay of organisms in the water, or the actual skeletons of living things. So they fall into a gray area. They can be either organic or chemical in origin. Sometimes, it can be a combination. In the case of skeletal remains we include them in with organic rocks. They also tend to be the most fossil rich of all sedimentary rocks. These rocks are usually very hard. They can be differentiated from other rocks by an acid test. If a weak acid of hydrochloric acid or that found in vinegar, is placed on a carbonate rock, it will fizz and possibly bubble. Be advised that non-carbonate rocks can contain carbonates in the matrix or as lithic fragments. So an acid alone isn't a slam dunk way of determining if a rock is a carbonate. Evaporites (mostly anhydrites), such as gypsum, are types of chemical rock. Although rare in Illinois, they are present mostly in the dolostone and limestone in Northeastern Illinois and a few places in the Mississippian rocks of Southern Illinois. Evaporites are mostly present in the Silurian and Devonian (p. 28 and 37).

Non-carbonate organic rocks are present in Illinois as coal. Illinois has vast abundance coal resources. Coal is usually black in color and weathers in blocks. In Illinois, it often has a sulfuric smell to it (Volume II, p.55-56). It has been extensively mined in Illinois but, most of it is shipped outside of the United States because of its high sulfur content. During the initial influx of settlers to Illinois in the 19th century, many homesteads were built in areas where coal was easily accessible at the surface. Some of the foundations of these old homes are still visible in state parks along the Illinois, Kankakee, and Spoon Rivers. You can often find these old homesteads by looking for out of place flowers that have reverted to wild flowers in the middle of the forest.

All sedimentary rocks have varying degree of impurities in them. It is common to have shaley sandstones (wacke), sandy limestones, calcareous sandstones, silty coal, cherty limestone, or any other multitude of combinations. There are very few pure rock types in Illinois, but they do exist. The Saint Peter Formation's, Starved Rock Member is 95% to 97% quartz, which makes it ideal for a source of groundwater and silica sand. The Racine Formation is pure dolostone in its ancient reef deposits, and is extensively quarried in Northeastern Illinois. Many Mississippian rocks contain high purity limestone.

Sedimentary rocks are also colored by certain minerals and elements. Red and reddish brown colored sedimentary rocks have significant iron concentrations. Green sedimentary rocks have abundant glauconite. Black or dark brown sedimentary rocks are usually high in organic content. Blue or gray hues generally indicate the presence of carbonates in sedimentary rocks. Color in sedimentary rocks can indicate the presence of a mineral, but it cannot be used to identify a rock. There is red dolostone in Northeastern Illinois. There are also red shales and red sandstones. There are Cambrian green sandstones but there is also green Ordovician shale. Almost all coal is black but there are also black limestone nodules in the Pennsylvanian deposits of Illinois. There is gray Silurian dolostone, but there are also gray Pennsylvanian rocks that are not carbonates. The color of a sedimentary rock is an indication of minor minerals within the rock. By itself, color is largely irrelevant when classifying sedimentary rocks.

Geologists treat all sedimentary rocks or loose sediment equally. We name rock or sediment units based on the lithology or type of rock. We do not care if the rock is sediment (loose deposits) or lithified (hard rock). This is spelled out in detail in a the document used for naming natural earth materials called the "North American Stratigraphic Code". The reason for this is simple. A rock unit can be lithified in one area or very soft in another. The Pearl Formation is exposed along the Illinois River (Volume II, p.64). In outcrop it is somewhat hard and you need a hammer to loosen it. Well logs from nearby indicate that it is softer underground. Yet, whether it is at the surface or underground it is all a conglomerate or gravel. If we used hardness as a means for classifying rock, things would get extremely complicated and messy. Instead of one rock unit we could end up with a dozen in a very small area. The hardness or lithification of a rock can change in short periods of time depending caused by natural forces (mineral precipitation or dissolution) or manmade mechanisms (alteration of groundwater chemistry or mining activity).

SEDIMENTARY ROCKS OF ILLINOIS		
NAME	DESCRIPTION	DEPOSITIONAL ENVIRONMENT
Anhydrite	White sulfate with calcium, it is deposited in a similar manner as other evaporites such as salt and gypsum.	Very shallow marine deposits in a restricted basin
Coal	Black, platy rock formed from compressed plant material, burns when heated.	Near shore swamps to non marine
Chert	In the rocks of Illinois it occurs in carbonate beds, precipitated out of sea water by micro and macro-organisms, comprised of silica. Occasionally fossils are present.	Shallow marine shelf
Erratics	Boulders and cobbles of either sedimentary, igneous, or metamorphic rocks, transported from a great distance by ice.	Boulders and cobbles in Glacial Diamicton
Limestone	A calcium carbonate that is deposited by sea water and living organisms. Fossils, oolites, and breccias are common. Usually varying shades of gray to brown. Red when iron is present.	Shallow marine shelf and freshwater lakes
Dolostone	A carbonate rock, similar to limestone in appearance and fossil content. Except it usually forms as groundwater transports magnesium into limestone altering its composition. It rarely is deposited directly from sea water unless brackish water is present.	Shallow marine shelf and the alteration of limestone through groundwater movement
Sandstone	A clastic rock that forms from the mechanical weathering of other rocks. Usually composed of quartz but can contain feldspar and lithic grains. Generally yellowish brown in color but can be white, gray, orange, red, or green if glauconite is present. Fossils are rare.	Near shore marine, marine shelf, rivers, lakes, glacial outwash, beach, dune deposits
Shale	A clastic rock that forms from the alteration of rocks into clay. If the clay is silty or sandy, it is usually called a mudstone. It can be almost any color. Grays and greenish grays are the most common in Illinois. Fossils are common.	Mudflats, marine shelf, deep marine, deltas, and rivers

IGNEOUS ROCKS OF ILLINOIS		
NAME	DESCRIPTION	DEPOSITIONAL ENVIRONMENT
Diabase	Permian aged intrusions in Southeastern Illinois. Composed of brown and gray peridotite, carbonatite, alnoite, and lamprophyre. Rare earth elements are known to present in the intrusions.	Intrusive igneous dikes, diatremes, small stocks, vent breccia
Erratics	Boulders and cobbles of either sedimentary, igneous, or metamorphic rocks, transported from a great distance by ice.	Boulders and cobbles in glacial diamicton
Granite	Other than what is found in erratics, granite is deeply buried beneath the Phanerozoic rocks in Illinois. It is usually a coarse grained crystalline rock that is red or gray in color and is comprised of quartz, feldspars, and other minerals.	Deep intrusive rocks that form largely from cooled magma chambers.
Rhyolite	Other than what is found in erratics, rhyolite is deeply buried beneath the Phanerozoic rocks in Illinois. It is usually a fine grained rock that is similar in composition to granite. It is commonly red and dark gray.	Shallow extrusive igneous rocks, usually deposited on the surface as ash flows.

METAMORPHIC ROCKS OF ILLINOIS		
NAME	DESCRIPTION	DEPOSITIONAL ENVIRONMENT
Erratics	Boulders and cobbles of either sedimentary, igneous, or metamorphic rocks, transported from a great distance by ice. All metamorphic rocks at the surface in Illinois were transported by glaciers from Wisconsin, Minnesota, Michigan, and Ontario.	Boulders and cobbles in glacial diamicton

ROCKS OF ILLINOIS: *Sedimentary rocks dominate the state. However, metamorphic and igneous rocks do exist, mostly as erratics or dikes in Southern Illinois.*

Geologic Structures

The type of sedimentary rock that is deposited in an environment is dependent on many factors. The source area that the rock came from (for clastic rocks), the temperature and current direction of the sea (for chemical rocks), the elevation at which the rock was deposited, the plants and animals living at the time the rocks were deposited, and the local geologic structures that formed the topography on the land and in the sea basins.

Illinois has dozens of named geologic structures. Only a few affected the majority of the State. The largest is the Illinois Basin, also called the East Interior Basin, (p.12-13). It formed during the Paleozoic and allowed for the transgression (or rise) of the ocean onto the land. The Illinois Basin is roughly spoon shaped and trends north-northwest to south-southeast. It is bordered to the north by the Wisconsin and Kankakee Arches (which served as high areas) the Mississippi River Arch to the west, the Ozark Dome and Pascola Arch to the south, and the Eastern Shelf to the east.

Inside the Illinois Basin are several large and prominent structures. The Sangamon Arch, LaSalle Anticlinorium, and the Fairfield Basin (the deepest part of the Illinois Basin) are all part of the larger Illinois Basin. The most influential of these structures is the LaSalle Anticlinorium which stretches from Lee County to Wabash County. The LaSalle Anticlinorium is visible at the surface in LaSalle County and it serves as a major area for gas and oil storage. It is actually comprised of connected smaller monoclines, anticlines, and synclines. Along the axis (or crest) of the Anticline, sediments containing coal have been brought close to the surface. This makes mining easier. Strip mines are common along the crest of the LaSalle Anticlinorium.

A subtle unnamed monocline is present at the Menominee Section in Jo Daviess County, along US-20 (Volume II, p.15). Notice how the relatively flat beds suddenly dip down to the west (arrow on the left). The arrow on the right (east) is pointing to flat beds. Photo is looking north-northeast. See page 63 for a detailed description of this stop.

Illinois isn't just a series of arches and basins, several large faults also run through the state. Even if you know nothing about faults, you have probably heard of the New Madrid earthquakes of 1811 and 1812. February 7, 1812 saw the largest earthquake in the lower 48 states near Marston Missouri. It was greater than 8.0 on the Richter scale. These occurred along the New Madrid Rift System, just south of New Madrid Missouri. Almost all those earthquakes were within 10 miles of the Mississippi River. The area is still seismically active in Southern Illinois and serves as the location for most local earthquakes. Most earthquakes that occur cannot be felt. Yet they happen all the time. As we research this book, an earthquake in Southern Illinois registered as a 1.9 on the Richter scale and occurred nine miles south of Jonesboro on July 1st, 2012, and no one felt it. Anything less than a 2.0 on the Richter scale is generally not felt by people.

Dipping beds of the Saint Peter Sandstone along Devil's Backbone Road, about 1.6 miles west-southwest of Oregon in Ogle County. These beds are dipping northeast, towards the northern most extent of the Sandwich Fault Zone, which lies about 800 feet to the north, and locally trends northwest to southeast. Photo is looking east. See Volume II, page 6 for a detailed outcrop description.

Southern Illinois isn't the only place for large faults and earthquakes in Illinois. Northern Illinois has two large inactive fault zones. The largest is the Plum River Fault Zone (which enters Illinois from Iowa) in Carroll County and it terminates in Ogle County. The Plum River Fault Zone is 112 miles in length and consists of high angle faults. In places rocks are offset by as much as 270 feet (at the surface) and as much as 1,100 feet of displacement is present in the deeply buried Precambrian rocks. It began to form during either the Mississippian Period or Pennsylvanian Period. It probably became inactive in the Permian. As the Plum River Fault Zone ceased to be, another fault zone began to form, the Sandwich Fault Zone.

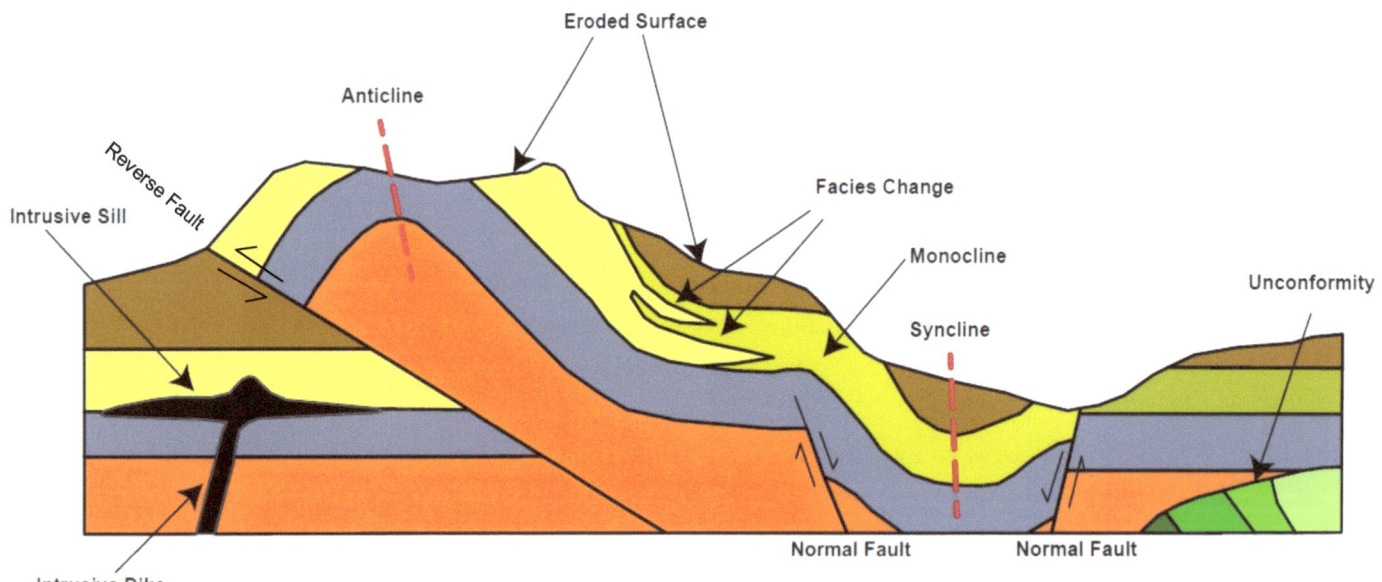

CROSS SECTION SHOWING GEOLOGIC STRUCTURES: *This is a hypothetical cross section or "slice" into the Earth depicting major geologic structures found in Illinois. The colored layers represent distinct geologic units.*

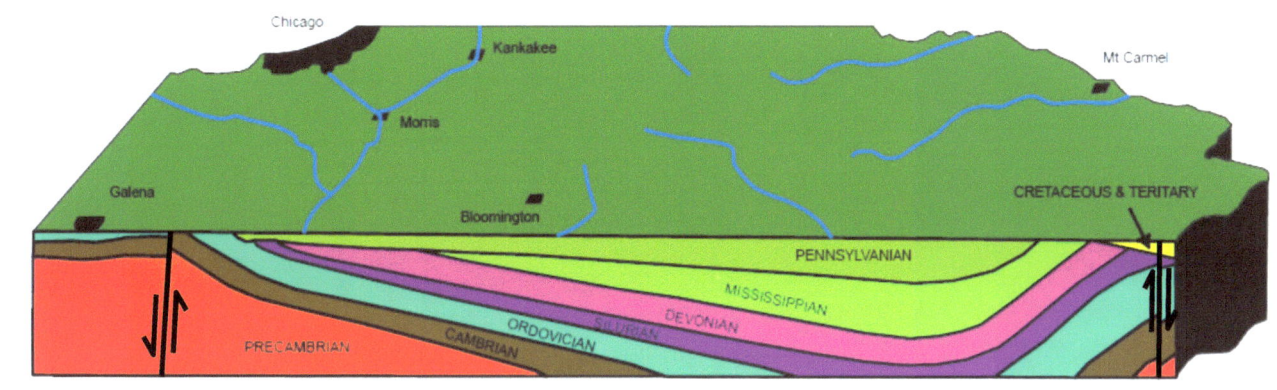
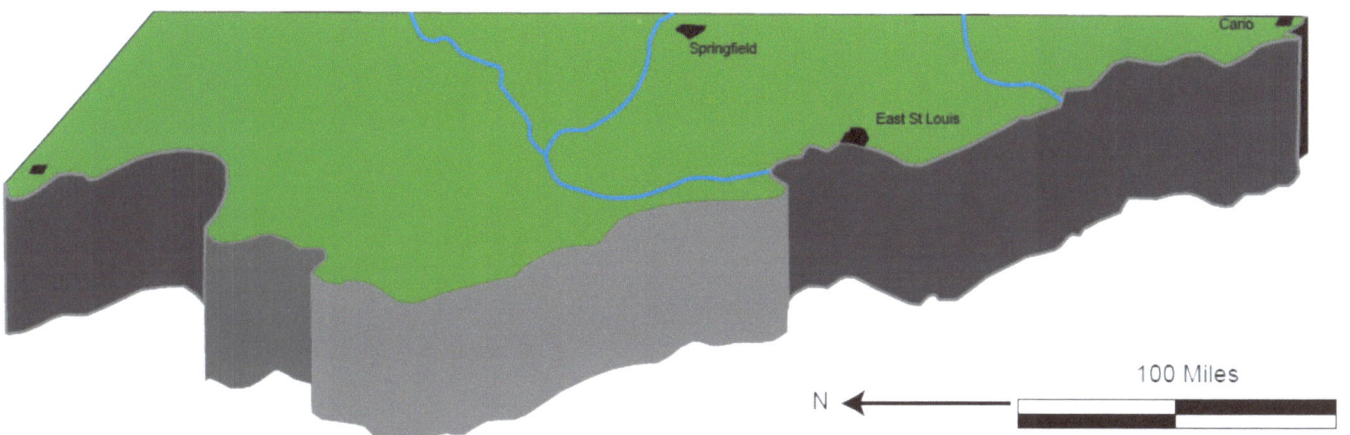

SLICE THROUGH ILLINOIS: *If we were to cut Illinois in half along a north to south axis we would see the major geologic periods represented in the rocks. Notice how they form a syncline near the center. The shape is a cross section through the Illinois Basin. The vertical lines with arrows offsetting geologic units in the north and south are major fault zones. Vertical exaggeration here is about 10 times.*

The Sandwich Fault Zone, like the Plum River Fault, is a linear and high angled fault zone. It begins in Ogle County, less than 10 miles from where the Plum River Fault Zone ends. In 2011 mapping was conducted by the Illinois State Geological Survey to determine if the Sandwich and Plum River Fault Zones connect. No field or subsurface evidence was found physically connecting the two fault zones.

The Sandwich Fault Zone extends continuously for 85 miles east-southeast into Will County. It is an average of two miles wide with rocks displaced from 200 feet to 800 feet vertically. The Sandwich Fault Zone is only exposed at the surface, where it begins near Oregon Illinois and where it ends just past Channahon Illinois. Elsewhere it is covered by glacial deposits. The age of the Sandwich Fault Zone is not exactly known. The youngest rocks faulted are Silurian and the overlying glacial deposits are not faulted. It could have formed any time between 415 million years ago and 50,000 years ago. Even with no direct evidence, there are other clues as to its age. The faulted rocks in Vicks Pit, east of Channahon, have sharp drag folds that formed near fault planes. The fact that these sediments are deformed, and not jumbled, indicate that the rock was already hard and buried when the fault zone formed. Another clue is the complexity of the fault zone. All three major fault types are present within the fault zone. There are normal, reverse, and strike-slip faults. This indicates several events of faulting with extension and compression from different directions. The general tectonic setting of Illinois indicates that it most likely formed during the Pennsylvanian through the Triassic, during the coalescing of the supercontinent Pangea, and possibly being reactivated later throughout the Jurassic and Cretaceous as Pangea broke up.

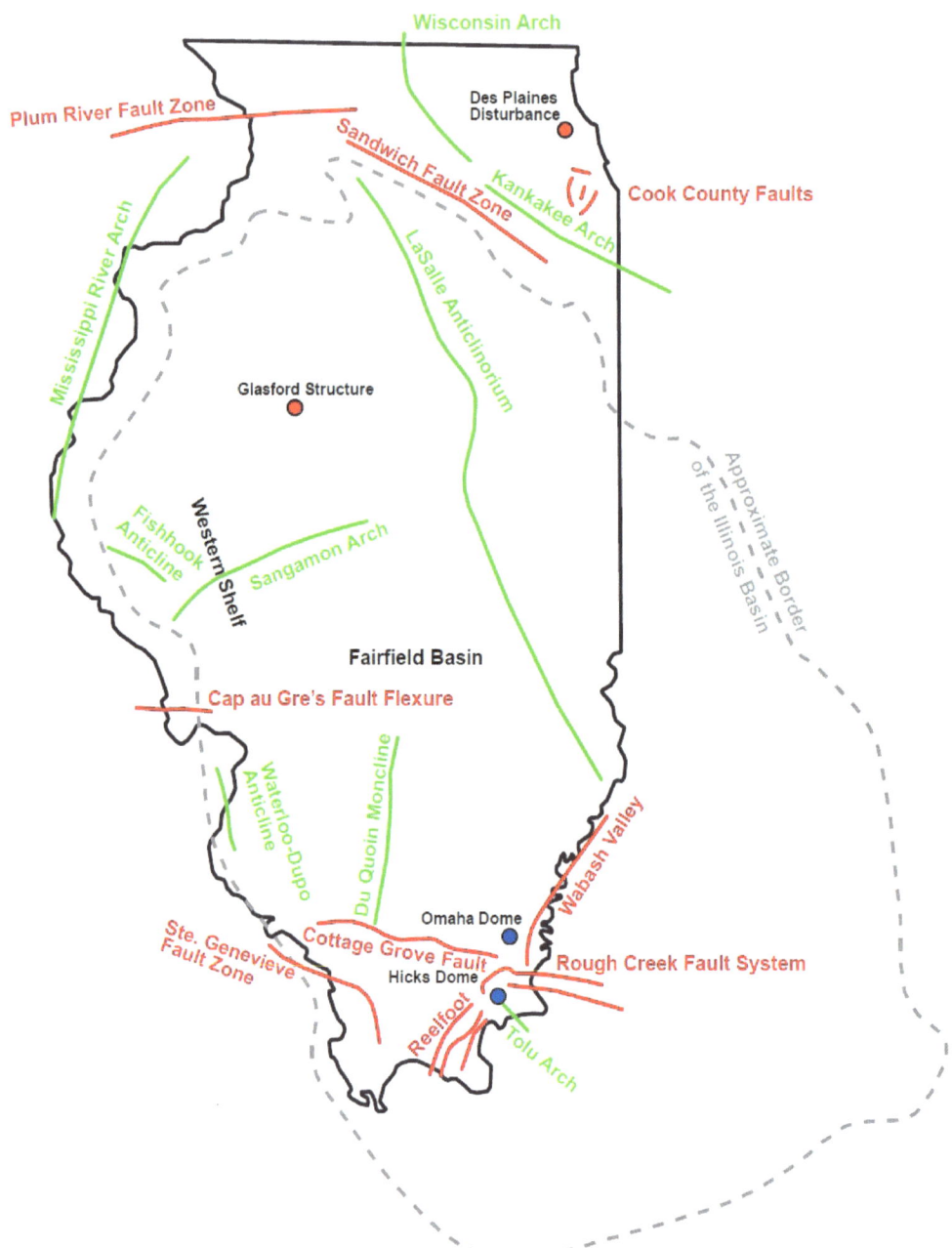

STRUCTURAL FEATURES IN ILLINOIS: *This map shows the location of the major geologic structures mentioned in this book. Red lines represent faults. Green lines represent the axis of folds. Red circles show the location of meteor impacts. Blue circles are crypto-volcanic structures. "Western Shelf" is a major platform within the Illinois Basin. The Fairfield Basin is the deepest part of the Illinois Basin.*

Earthquakes also occur in Northern Illinois. If you plot them on a map, (such as the one that occurred in Kane County in February 2010) they don't fall along the Sandwich or Plum River Fault Zones. This is somewhat of an enigma because faults are the logical place for earthquakes to occur. They also originate deep with the Precambrian rocks miles below the surface. The Precambrian rocks are poorly understood in Illinois since none of them are at the surface. Some earthquakes do occur along geologic structures. A large number of the earthquakes in North-Central Illinois occur on the LaSalle Anticlinorium. This indicates that the anticlinorium is active again. The focus (the actual point within the earth that an earthquake occurs) of most of the earthquakes in the northern half of Illinois occur miles underground.

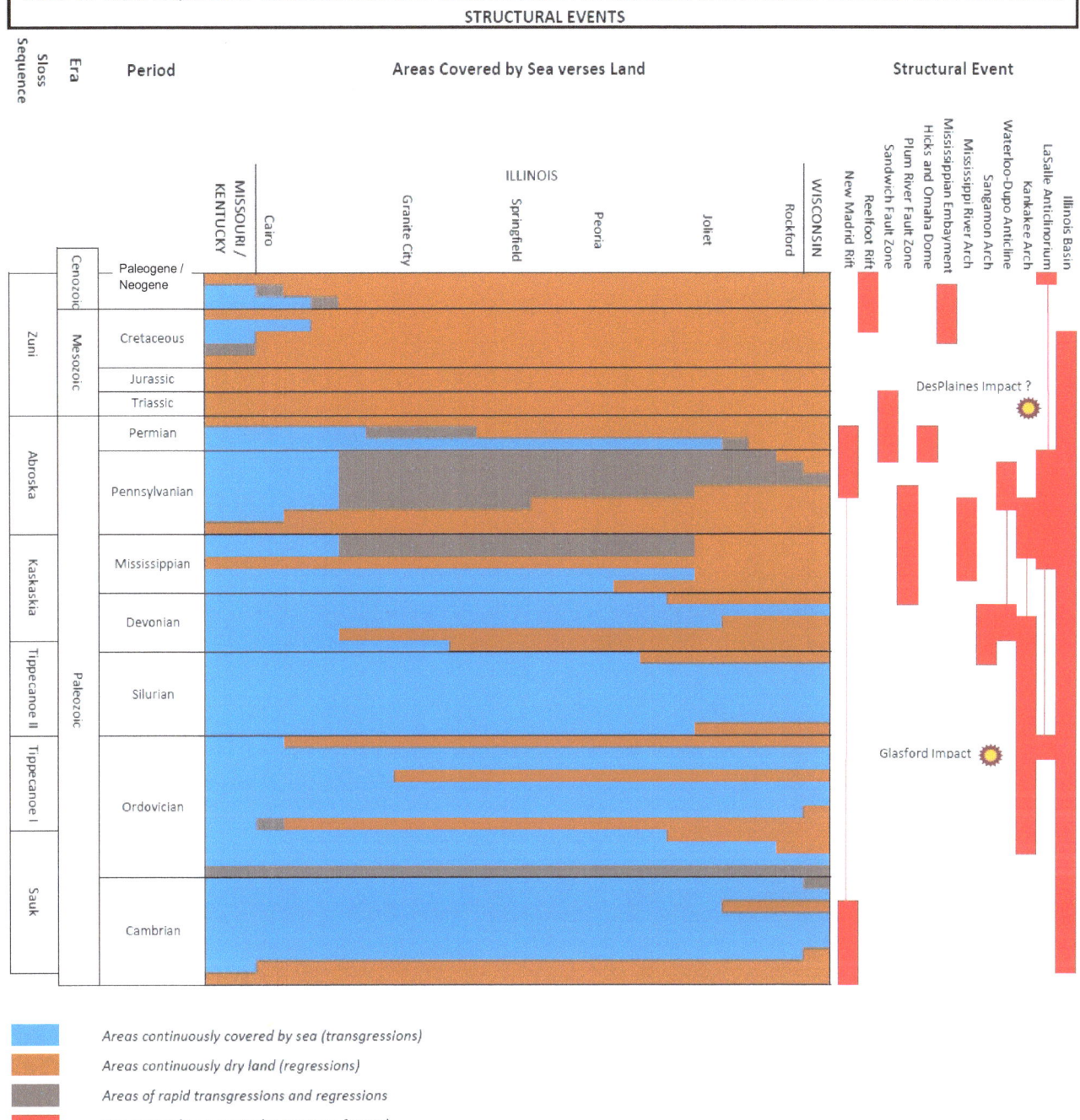

Geologic Time and Formations

Geologists have divided the history of the Earth into named segments of time, with divisions occurring along major geologic events, change in fauna, or rock sequences. The further back you go in time the larger the time spans included in a geologic time period. As you back further in time, the rock record becomes less certain because less rocks are preserved.

Geologic time is divided into (from smallest to largest) ages, epochs, periods, eras, and eons. There are some divisions of time that are further clumped together or divided apart. The Precambrian is a supereon. This is a holdover from the 19th century when the Precambrian was thought to represent a period of time as long as the Phanerozoic Eon (541 million years ago to the present). We now know that the Precambrian contains at least 88% of the entire span of time from the beginning of Earth to the present. There are Precambrian sedimentary rock formations in places (such as in Ontario, Montana, and the Grand Canyon) that are vastly thicker than the entire sequence of Phanerozoic sequence of rocks in Illinois.

Although geologic time is important to piecing together the history of the Earth, it is independent of how geologist name and classify rock units (Specifically lithostratigraphic units). Geologists organize and name rock units based on similar characteristics and physical properties.

The division of geology pertaining to the naming of rocks is called stratigraphy. We use lithostratigraphy to map geologic units. Rocks of similar type that occur in at the same levels in the rock record are given formal names. In lithostratigraphy, the unit basic unit of rock is called a formation. And the name of a formal rock unit will usually have the word "Formation" as part of its name. For example, the most extensive surficial glacial diamicton in northeastern Illinois is the Lemont Formation. When a geologist talks about the Saint Peter Formation (also known as the Saint Peter Sandstone), we are talking about an assemblage of rocks, which are made of a similar rock type independent of time. Formations can cover several states or be restricted to a single county. They can be miles thick or tens of feet thick.

There are three basic criteria for defining a formation. It must be easily identifiable, mappable, and named after a semi-permanent natural or manmade feature (i.e. rivers, hill tops, mountain passes, towns, etc.). Formations cannot be named after people or short-lived geographical features. Formations can be divided into formal subunits such as members, lenses, and tongues, if there is an economic or scientific reason for doing so. Formations can also be parts of groups or supergroups. For example, the Starved Rock Member is part of the Saint Peter Formation, which belongs to the Ancell Group. Yet, not all formations are subdivided, nor do all belong to groups. Confused? Well, geologists like all scientists, like to name things. It is a bit overwhelming at first, but you get used to it.

STRATIGRAPHIC RELATIONSHIPS OF CAMBRIAN AND SOME ORDOVICIAN ROCKS IN NORTHERN ILLINOIS						
Lithological Units				Time Units		
Supergroup	Group	Formation	Member	Period	Epoch	Mya
	Ancell	Glenwood		Ordovician	Middle	458
		Saint Peter	Starved Rock			
			Tonti			
			Kress			470
Knox		Everton			Early	
	Prairie du Chien	Shakopee				
		New Richmond				
		Oneota	Blodgett			
			Arsenal			485
		Jordan		Cambrian	Furongian	
		Eminence				
		Potosi				497
Potsdam		Franconia	Derby-Doerun		Epoch 3	
			Davis			
		Ironton				509
		Galesville				
		Eau Claire	Proviso		Epoch 2	
			Lombard			
			Elmhurst			
		Mount Simon				521
					Terreneuvian	541

Mya = millions of years ago

Chart shows how lithologic units can cross time boundaries

Not all named units are listed

Time scale is adapted from the GSA 2012 Geologic Time Scale

Geomorphology and Landforms

The study of the landscape and its features is called geomorphology. Ridge lines, river courses, lake locations, vegetative cover, and even how we use the land are all clues as to the geology. After all, the landscape is a function of the geology, not the other way around. Illinois looks the way it does because of more than 1.6 billion years of geologic events ranging from Precambrian volcanos right up to the ice ages of today. If any single geologic event did not occur when it did in Illinois, the landscape would be very different. If there were no hard carbonate Silurian deposits in the Chicagoland area, Chicago would not exist because the Ordovician sediments would not support such massive skyscrapers. If Illinois was significantly above sea level 315 million years ago, we wouldn't have any coal. If the ice sheets of the past one million years had not covered Illinois, we would not have the vast fertile farmlands of today, nor would Lake Michigan exist. The opposite is also true. Say the New Madrid Rift Zone had fully developed and split the continent. Illinois would not be at the longitude and latitude that it is today. Who knows where it could have ended up?

Fortunately Illinois did not split from the rest of North America, the seas did cover the Illinois throughout most of the Paleozoic, and the glaciers did advance and retreat. All these events are recorded in the rocks and shaped the landscape of today. We use geomorphology to divide Illinois into physiographic areas (p.19).

There are several significant landforms in Illinois. Although they aren't anything as grand as faults, they directly impact human activity and fall under geologic hazards. Some are natural and some are manmade. Perhaps the most common of the smaller natural landforms are sinkholes. Most sinkholes form in karst areas. Karst areas are most common where the glacial deposits are absent or thin, and the bedrock near the surface is carbonate rock. They are common along the Mississippi, Ohio, and Wabash Rivers of Southern and Western Illinois. Karst areas are beautiful and are where many cave systems are found, but karst areas are also prone to sinkholes. Sinkholes form when underlying carbonate is dissolved away through groundwater activity forming a small underground cave. The soil or thin loess material above becomes unstable and collapses into the cave forming a cone structure at the surface. Sinkholes can be anywhere from a few feet deep to dozens of feet deep. They can form in a matter of seconds!

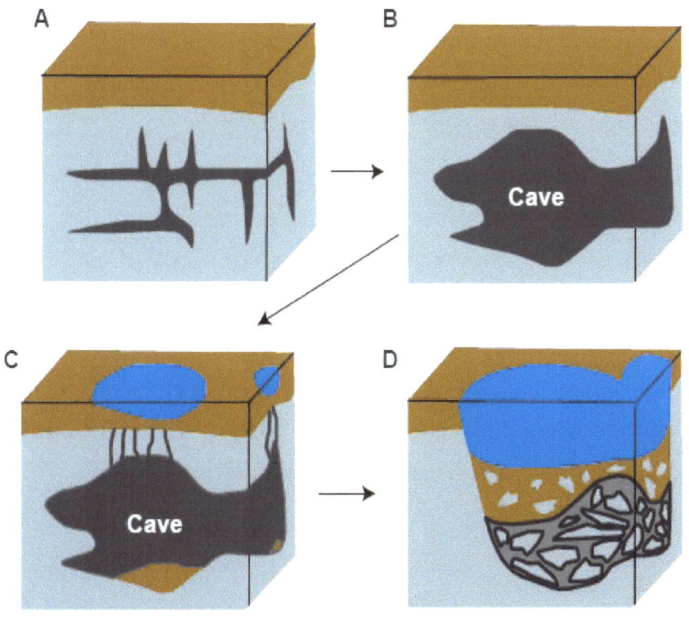

KARST FORMATION: *This diagram shows how limestone becomes part of the karst topography and ultimately a sinkhole. A) Shows how limestone dissolves along joints in the subsurface due to enriched groundwater. B) Eventually the small openings form a large cave. C) A small sinkhole can form when soil from the surface falls into the cave. D) A large sinkhole can form from the complete collapse of the subterranean cave system.*

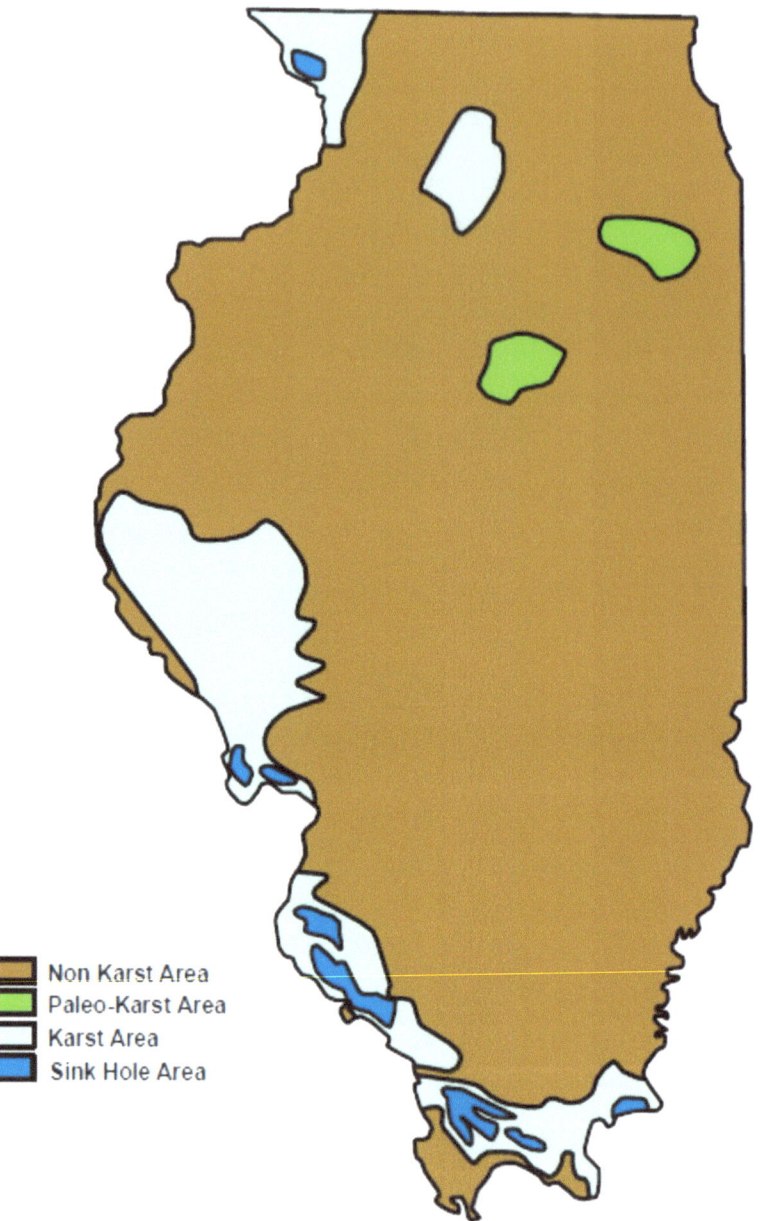

KARST LOCATIONS IN ILLINOIS:
This generalized map of Illinois shows where the main karst and sinkhole areas are located. Paleo-karsts are caves and sinkholes that have formed or been filled in with sediment prior to the Quaternary. They also exist in the main karst areas but are too small to show up on the map.

- Non Karst Area
- Paleo-Karst Area
- Karst Area
- Sink Hole Area

There are also manmade sinkholes outside of karst areas. In central Illinois old underground coal mines often collapse forming a type of sinkhole. In Northeastern Illinois, on the Chicago Lake Plain, most people are familiar with potholes. There are two types of potholes. The first type is those caused by surface wear and tear. The second are more dangerous. Anyone driving on the extreme North or South Side near the lake is familiar with potholes that are only six inches to two feet wide. If you look down into them, you can't see the bottom. These are caused when underground storm sewers collapse or water mains fail in sandy areas. One such large manmade sinkhole occurred near the 9600 block of South Houston Street in Chicago on April 18th, 2013. At about 5AM, a sinkhole the width of the street opened up and swallowed three vehicles. It formed when a waterline built in 1915, burst underground and spilled into the underlying sewer. This, along with the heavy rains in April weakened and undermined the soil above. The soil was carried away in the storm sewer causing a sinkhole to form.

Rivers play an important part in shaping Illinois. As rivers become established they stop eroding down and start to erode laterally (from side to side). As a result, wide valleys form called floodplains. Rivers will move laterally overtime within the floodplain, and as they do the river channels curve and bow into meanders. Although major rivers can exist for millions of years, individual meanders are short lived. They will often get "cut off" from the main river channel as the river finds an easier path to take. These cutoffs often become oxbow lakes until they dry up. Rivers, no matter how large or small will naturally do this (Volume II, p.58). Small streams like those common in Western Illinois, can form cutoffs in a matter of years. Other large rivers like the Mississippi River will form cutoffs in a matter of decades or centuries. We can see these processes reflected in the stateliness. One town in Western Illinois called Kaskaskia (Randolph County) was cut off from the rest of Illinois in 1881 due to flooding caused by deforestation. Once the flood waters subsided, the Mississippi River moved to the east essentially dropping Kaskaskia on the Missouri side! Kaskaskia officially stayed a part of Illinois, and a levee was built to the east of the town.

Manmade levees are built along rivers in a futile attempt to keep a river in its channel. Levees are poorly planned structures, no matter how well they are built. Unlike natural levees which form during flood events, manmade ones are built to stop flooding. Manmade levees are usually built in straight ridges along a river. This forces the river to do something very unnatural. It forces it to flow in a straight line. This increases the water velocity and height of the river during flooding. Established rivers do not like to flow in straight lines. They slowly erode the sides and bases of levees. All it takes is one major flood event to break them down. This happened in Illinois along the Mississippi River during the floods of 1993, then again along the Ohio and Wabash Rivers in 2011.

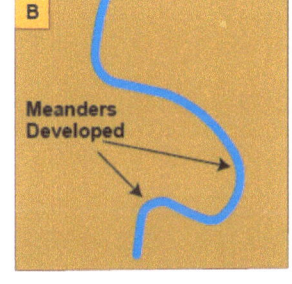
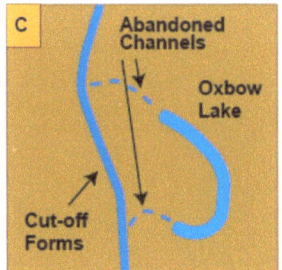
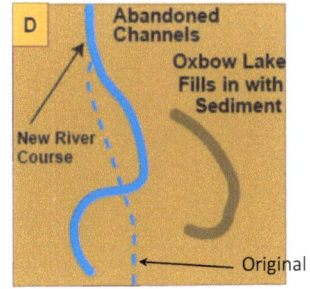

RIVER MEANDER DEVELOPMENT: *Once a river becomes established, it never follows a straight course. The above diagram is a map view of how rivers change with time. A) Depicts a river before it begins to form meanders. B) Meanders develop as the river starts to erode horizontally. C) An oxbow lake can form after a river changes its course, usually after a major flood events. D) The river will continue to change its course, long after the oxbow lake has filled in with sediment.*

Another beautiful landform in Illinois is the lakes. Most natural lakes in Illinois were left by the glaciers as they retreated. The Chain O'Lakes area in McHenry and Lake Counties formed when ice trapped under glacial till melted and formed depressions. These water filled depressions are called kettle lakes. Kettle lakes are short lived and usually shallow. Over time they fill in to become bogs and eventually completely fill in with sediment. Other natural lakes are left directly by the melting glaciers such as Lake Michigan. There have been other large lakes in Illinois that are no longer around. They are called ice dammed or proglacial lakes. These lakes form when glacial melt water becomes dammed up by an end moraine or an esker. When the natural dam fails they catastrophically empty, leaving behind flat areas. Such lakes existed in Grundy, Kankakee, Iroquois, and Will Counties.

Illinois has been shaped by water, wind, fire, and ice. Illinois would be a very different place if it weren't for the volcanos 1.6 billion years ago, the failed rifts one billion years ago, the seas 300 million years ago, the ice and wind of 25,000 years ago, or the rivers of today. All of these natural forces have contributed to the landscape of the Prairie State over time. These divisions are described in detail in Volume II, pages 5, 23, 49, 62, and 71.

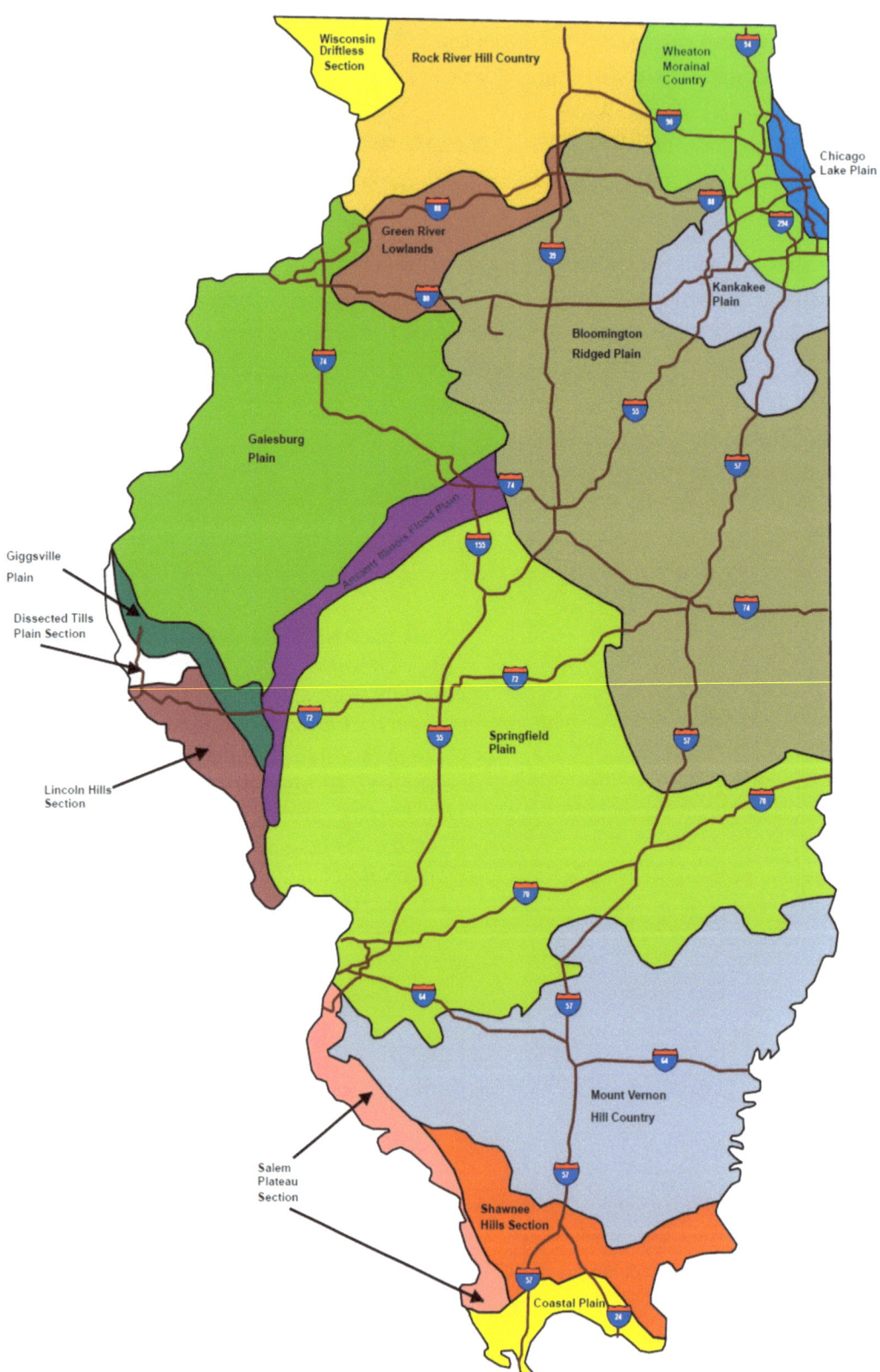

PHYSIOGRAPHIC DIVISIONS: *Illinois can be divided into at least 17 physiographic divisions based on geography, land forms, and the environment. These physiographic provinces are the basis for dividing Volume II.*

Introduction to the Physical History of the Prairie State

Mesoproterozoic (1.6 to 1.0 billion years ago) and Neoproterozoic (1.0 billion years to 541 million years ago) Eras: Illinois' Roots and the Great Unconformity

Geologically speaking, Illinois is physically very young. There are no rocks at the surface older than about 500 million years. The oldest and deepest rocks in Illinois that have been dated only go back to 1.6 billion years to the Mesoproterozoic Era of the Precambrian and consist almost exclusively of deeply buried igneous rocks of granite and rhyolite. That isn't to say that there are not even older and deeper buried rocks. They just haven't been found.

These deeply buried rocks are referred to as "basement rocks" because they occur below the sedimentary rocks in Illinois. The first 2.9 billion years of Earth's history (roughly the first two thirds), there either was no land to form Illinois or the rocks were eroded long ago. Chemical and mechanical weathering (mechanisms of erosion) was the dominant force shaping the land in Illinois during the Precambrian. There were no land plants or animals during this time and most life was still single celled and lived in the oceans. With no plants to anchor the surface deposits, erosion worked at a faster pace than it does today.

In Illinois the oldest rocks are deeply buried beneath thousands of feet of younger rocks. The closest these rocks come to the surface is about 1,500 feet below the surface in an east-west trend from Jo Daviess to Boone Counties, adjacent to the Sandwich Fault Zone.

Location: *Display case at the Illinois State Geological Survey. This is a cut and polished core of Precambrian red granite. This is from one of the rare cores taken from Central Illinois.*

In southern Illinois the oldest rocks are at the deepest in places in Saline, Hamilton, and White Counties (where they are 14,000 feet below the surface). The entire extent of the basement rocks in Illinois have been assigned to the Centralia Sequence divided into several subsequences, some of the subsequences may be sedimentary in origin. We just don't know the extent and variety of Precambrian rocks in Illinois, since most have never been drilled into. Almost everything we know about the Precambrian in Illinois is data from seismic profiling.

Through seismic profiling we can determine a basic geologic setting during the Mesoproterozoic and there was a lot of igneous activity. In eastern Illinois, we don't get the high magnetic and high gravity readings typical of a rift zone, like the rift under Lake Superior. The best alternate explanation is the basement rocks formed in a manner similar to the Yellowstone Caldera today. A "hot spot" of magma moves up through the Earth's crust and forms a large super volcano about 1.6 billion years ago. This also occurred in the area of the Ozarks in Missouri, around 1.45 billion years ago. These hot spots can be active for tens of millions of years and stay fixed relative to the interior of the earth, but appear to move as the continent passes over them. This event may also be part of a large continental wide hydrothermal event that affected Precambrian rocks in Wisconsin, Minnesota, and Michigan (the Wolf River Event) around 1.4 billion years ago. Once all of the upwelling magma was released in Illinois during the Mesoproterozoic, the land began to subside and Illinois' largest geologic structure, the Illinois Basin, began to form.

Illinois was essentially stable and quiet again until 880 million years ago during the Neoproterozoic, with exception of the formation of the Reelfoot Rift and the break-up of an older supercontinent called Pannotia. The Reelfoot is a failed rift system. The continent began to split apart but then stopped. May different things can cause a rift to fail. The most common way rifts fail involves a collision with another continent or the magma plume feeding the rift isn't large enough. Once the rift failed, subsidence of the Illinois basin accelerated and around 535 million years ago, the oceans began to cover all of Illinois, mainly from the south.

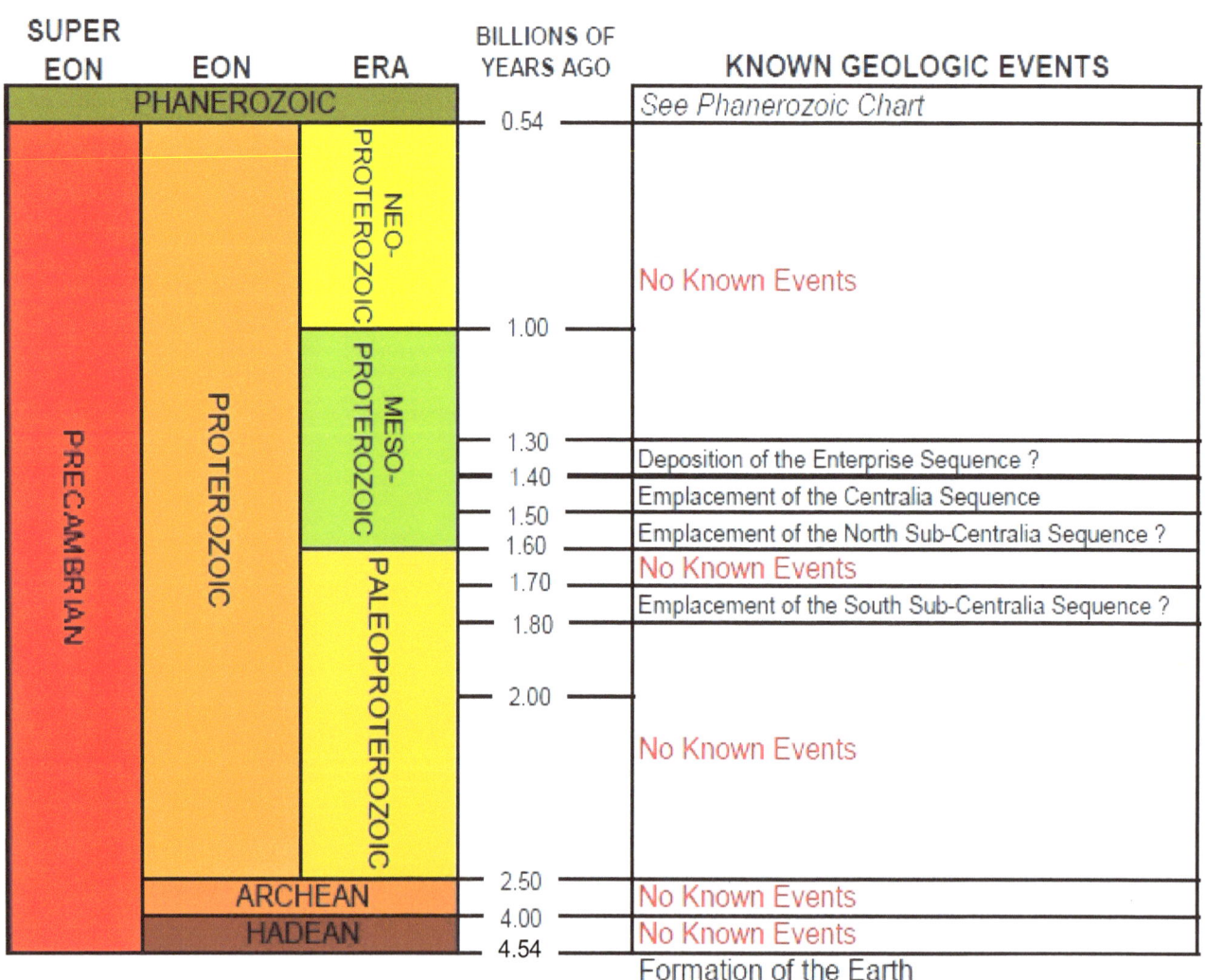

Phanerozoic Eon, Paleozoic Era: Appearance of Vast Marine Life

Cambrian Period (541 to 485million years ago): Explosion of Life

The Cambrian is separated from the underlying Precambrian rocks by "The Great Unconformity". Unconformities are noticeable gaps in the rock record where rock was either not deposited or has been eroded. In Illinois the Great Unconformity represents 850 to 870 million years of missing time! This is over one and a half times longer than from the Cambrian to today (or the Phanerozoic Eon).

The Cambrian ushered in a new era for Illinois. Long gone are the super volcanos and most of the rifts, although the New Madrid Rift in Southern Illinois would remain active through the first half of the Ordovician before becoming "reactivated" in the Pennsylvanian, and in modern times. The Cambrian saw the beginning of a long period of relative tectonic serenity and vast shallow seas that would persist, with only minor interruptions, for more than 300 million years.

The rocks from this time are all sedimentary and in the lower part are almost all are sandstones separated by the carbonates and shales of the Eau Claire Formation (Northern Illinois) or the Bonnterre Formation (Southern Illinois). The rocks deposited by the covering of the initial shallow oceans, is referred to as the Sauk Sequence. A sedimentary sequence is a series of rocks that are deposited in a predictable order on shallow ocean platforms. The upper part of the Cambrian is mostly carbonate rocks in Illinois that become sandier as you go north from southern Illinois. By the time you reach the northern border of Illinois, most of the Cambrian rocks are sandstones capped with carbonates.

Most of the Cambrian sandstones in Illinois contain at least some glauconite. Glauconite is a soft green to dark green mineral formed in shallow seas as a product of microorganisms interacting with seawater and other organisms near the shore. The Franconia Formation in Illinois contains so much glauconite that drillers call it "green sand".

The Cambrian is also the first time that large multi-cellular life appears in abundance in the fossil record. Although many of the animals and plants alive in the Cambrian seas may look exotic, they were closely related to modern forms of life today. Some would thrive for millions of years before becoming extinct. Others would survive into modern times.

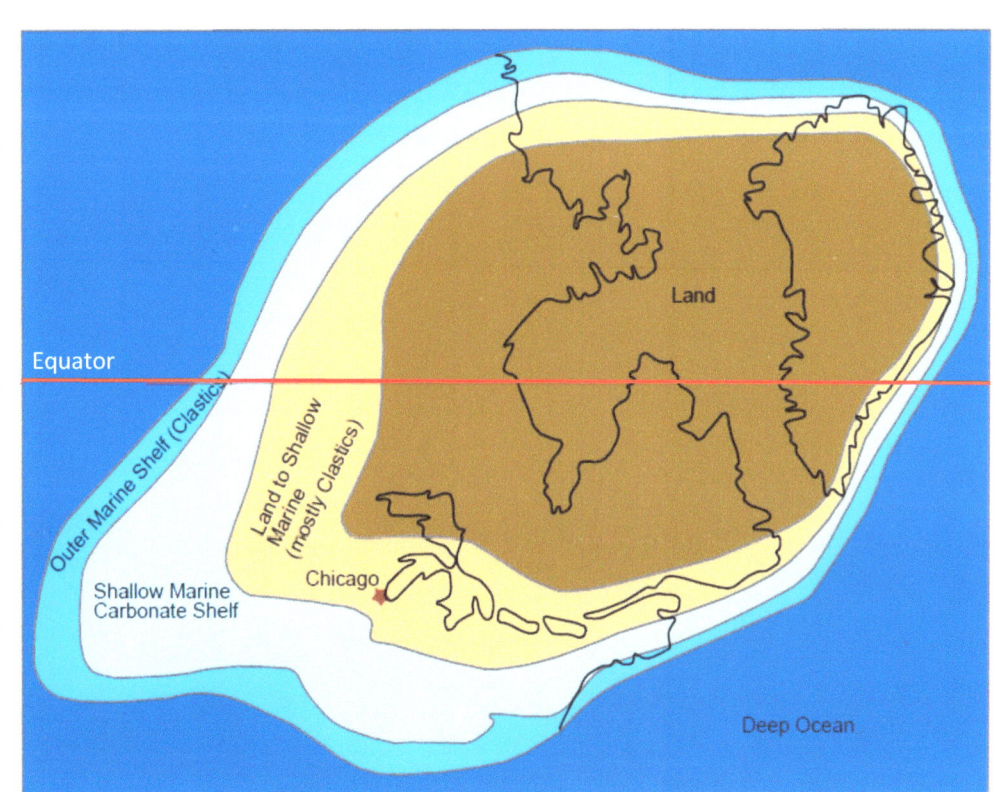

CAMBRIAN PALEOGEOGRAPHY: 520 million years ago, North America was in a very different location then it is today. During the Cambrian, North America was at the equator and Illinois was south of it. The continent was also smaller. The Rocky Mountains and most of the South wouldn't begin to form for almost another 400 million years.

The landscape of Illinois during the Cambrian was very different than what we see today. All of Illinois was covered by shallow seas, with only brief low lying islands emerging from time to time as sea levels would or fall (regress) before rising (transgress) again. Although the Cambrian seas were teeming with life, the land was devoid of any plants or animals except for bacteria and possibly fungi.

Fossils in the Cambrian of Illinois are rare, since so little of the Cambrian is exposed at the surface. Mineralization is very common. Color banded chert and small quartz filled geodes are common in Northern Illinois.

Location: Ogle County Illinois. The photo shows small oval geodes filled with clear quartz.

Location: Ogle County Illinois. The photo shows the brown color banding in Cambrian chert. Most chert in Illinois is white or gray. Some browns and yellows are common in the Cambrian and Lower Ordovician rocks of Illinois, but are generally lacking in Middle Ordovician-Devonian chert.

Ordovician Period (485 to 444 million years ago): Time of Crinoids, Brachiopods, and Glaciers

Geologically, the Ordovician was an active time throughout Illinois. The Sauk Seas fully retreated from Illinois and the surrounding areas about halfway through the Ordovician, before returning in the second half. It is during this time that erosion resumed. The seas remained just south of Illinois so the unconformity during this time is less in Southern Illinois (about 1 million years) and greater in Northern Illinois (about 8 million years).

At this time sea life spread like wild fire, crinoids were one of the dominant animals in the oceans along with bryozoans, brachiopods, and trilobites. Trilobites are now extinct, but the horseshoe crab is a close relative. Brachiopods are still around and numerous, although their diversity has declined since the end of the Paleozoic. Crinoids are also still alive today (although a lot rarer) and are referred to as "sea lilies". This name is misleading since they are actually animals and not plants. Almost all ancient crinoids had stalks and were anchored to the sea floor by a foot resembling roots. Most modern crinoids are free swimmers, but some with stalks still exist. During the early part of the Ordovician fish become abundant but only jawless varieties existed.

Location: Display case at the Illinois State Geological Survey. This cut and polished core from the Galena Group contains abundant bryozoans (moss animals). Bryozoans are still alive today. They live in both salt and freshwater. Here the fossils are somewhat broken but they make up most of the dolostone rock. Such a rock is referred to as a skeletal packstone.

Location: Display case at the Illinois State Geological Survey. Locally the Shakopee Formation is the formation just below the Ancell Group. It was highly eroded soon after deposition. This polished core from Central Illinois shows the wavy nature of the fine grained part of the formation.

When the seas returned during the middle of the Ordovician, a new sequence of rocks began to be deposited and is collectively referred to as the Tippecanoe Sequence. During this time carbonate rock deposition all but stopped. Deposition continued with the Saint Peter Formation and its associated rocks of the Ancell Group. The Saint Peter is one of the purest quartz sandstones in the world and also one of the best sorted. The Saint Peter is mostly a fine to medium grained, well rounded, quartz sandstone. As a result of its purity and thickness (generally 100 to 450 feet thick), it is a major source of groundwater for much of the State. In Illinois, the Saint Peter is represented by shallow marine sands and bars. However, in parts of Wisconsin and Missouri, it appears to be beach sand.

In Northern Illinois, the Saint Peter is capped with about 5 to 75 feet of the Glenwood Formation (part of the Ancell Group). The Glenwood is mostly impure sandstone. Certain parts of the Glenwood are referred to by drillers as the "re-peter sandstone" and were formed in a restricted lagoon environment. From LaSalle to Ogle Counties there is an unconformity on top of the Saint Peter. The exact amount of missing time is not known but it is probably only a few hundred thousand years. In Ogle County the unconformity is angular. The formation of this unconformity is probably due to the initial formation of the LaSalle Anticlinorium, the Kankakee Arch, and the Wisconsin Arch. South of LaSalle County, the top of the Ancell Group consists of the Joachim Dolostone and the Dutchtown Limestone. The Glenwood was never deposited south of LaSalle County. The Ancell Group demonstrates the typical lithology of a sedimentary sequence.

At the end of the deposition of the Ancell Group, the Illinois basin returned to a carbonate platform environment. It is during this time that the Galena and Platteville Groups were deposited. The Galena Group is equivalent to the Trenton Group outside of Illinois. Both the Galena and Platteville consist of mostly dolostone with minor amounts of limestone and shale. The two groups are separated by the shale and carbonates of the Decorah Formation. The uppermost member of the Decorah Formation is the Guttenberg Member in Northern Illinois. The Guttenberg is known to contain hydrocarbons and is often called "oil rock" by drillers. Minerals are extremely common in the Galena-Platteville of Northern Illinois. Lead and zinc is common but so are other, less economic yet beautiful minerals such as calcite and pyrite.

Location: Open display in the Illinois State Geological Survey Annex. This specimen was obtained from the Conco Mine in Aurora, Illinois, on the north side of I-88. Calcite, marcasite, and pyrite are present in this specimen obtained in the subsurface part of the mine from the Wise Lake Formation of the Galena Group.

The Galena and Platteville Groups also record a geologic event that was occurring several hundreds of miles to the east. During this time North America was colliding with Northern Europe. This created massive volcanoes. As they erupted, ash was kicked up into the air that would be carried west by the winds into Illinois and be deposited as thin beds of yellow bentonite. Bentonite is a volcanic ash made of highly expandable clay and is mined in places like Wyoming. When water is added to the dry clay it can increase in volume fivefold. It is used as a sealer in water wells.

At the end of the Ordovician, clastic deposition would once again resume, with the Maquoketa Group. Only this time shale and not sandstone would be the dominant rock deposited. The Maquoketa Group caps the Galena and Platteville Groups. At the very end of the Ordovician, the seas began to slowly retreat. First deep dark shale would be deposited followed by coarse fossil rich carbonate, followed by light colored shale. In northern Illinois patches of a deep red to purple iron rich formation exist. This iron formation is mined in Wisconsin but is too thin to be economical in Illinois. It is the Neda Formation and the iron is contained in abundant little black balls of hematite less than half a millimeter thick. These little iron balls or pellets care called oolites. Oolites are usually made of limestone or dolomite. In the Neda, they are mostly hematite (Volume II, p.41). The hematite oolites were most likely deposited as limestone but were later altered, by either exposure to the atmosphere or groundwater.

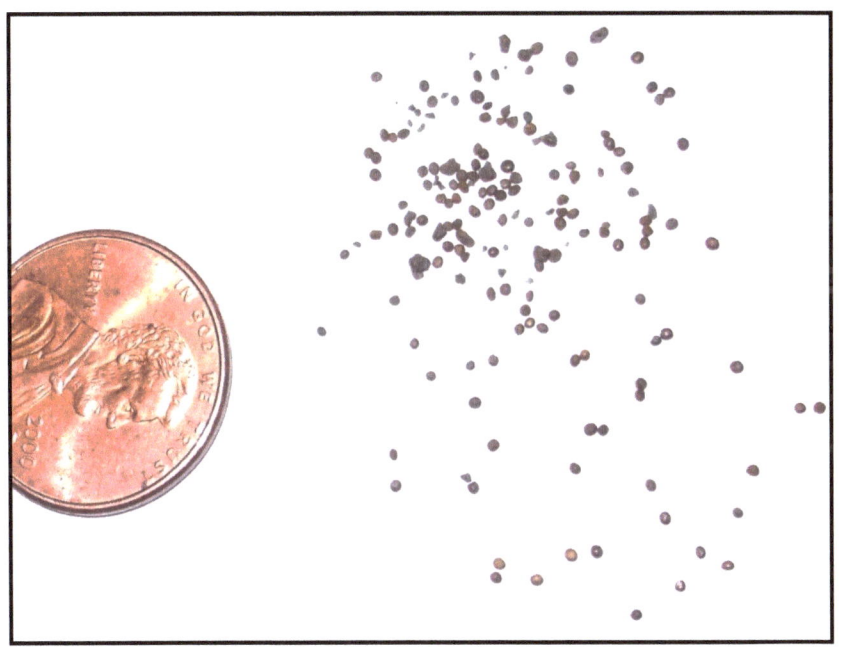

Individual iron oolites from the Neda Formation. Most of these round balls were originally deposited as calcite, which was later replaced with hematite.

Once the Neda was deposited, the seas rapidly withdrew and vigorous erosion occurred throughout Northern Illinois. As much as two thirds of the Maquoketa Group was eroded in places. The cause of this rapid drop in sea level was due to an ice age in the Southern Hemisphere, since there were no significant land plants to anchor the soil. During the end of the Ordovician, most of the Earth's land was near the South Pole. Illinois at the time was close to the equator and not directly covered in ice. Sea level dropped worldwide by as much as 200 to 300 feet exposing rocks to the surface. This Ordovician Ice Age probably only lasted about three to five million years.

Silurian (444 to 419 million years ago) and Devonian (419 to 359 million years ago) Periods: Great Reefs and Clear Seas

During the beginning of the Silurian, the Ordovician Ice Age ended and the seas once again covered the entire state. In the northern third of Illinois, the Silurian is mostly dolostone rock. In southern Illinois, limestone is more common. As the seas transgressed over the eroded Ordovician hills and valleys, the initial deposits were carbonate muds mixed with shale. Carbonates with large amounts of shale are referred to as argillaceous carbonates. As the Silurian deposits became thicker and thicker, they became less argillaceous and more fossil rich. The Silurian rocks in Illinois contain some of the best fossils in the world from this time period, especially in the reef rock.

Halysites sp. ("chain" coral)
New Baden East Reef
Silurian
Moccasin Springs Formation

Clinton County
Depth 1967 ft

Rugose coral
in crinoidal grainstone
New Baden East Reef
Silurian
Moccasin Springs Formation

Clinton County
Depth 1937 ft

Location: Display case at the Illinois State Geological Survey. The Silurian reefs of Central and Southern Illinois contain more limestone than dolostone. As a result fossils are far better preserved, as in these corals from rock cores drilled in Central Illinois.

There are two well defined unconformities in the Silurian that appear to be nearly worldwide. One of the two unconformities occurs within the Kankakee Formation and separates the Drummond and Offerman Members. This unconformity has only been recently recognized. The second unconformity separates the lower Kankakee Formation from the overlying Joliet Formation. This unconformity has been known for a more than a century. These two unconformities had a nearly global extent, yet the estimated amount of missing time is probably less than a million years for each. It is exceedingly rare for such global unconformities to represent such little missing time.

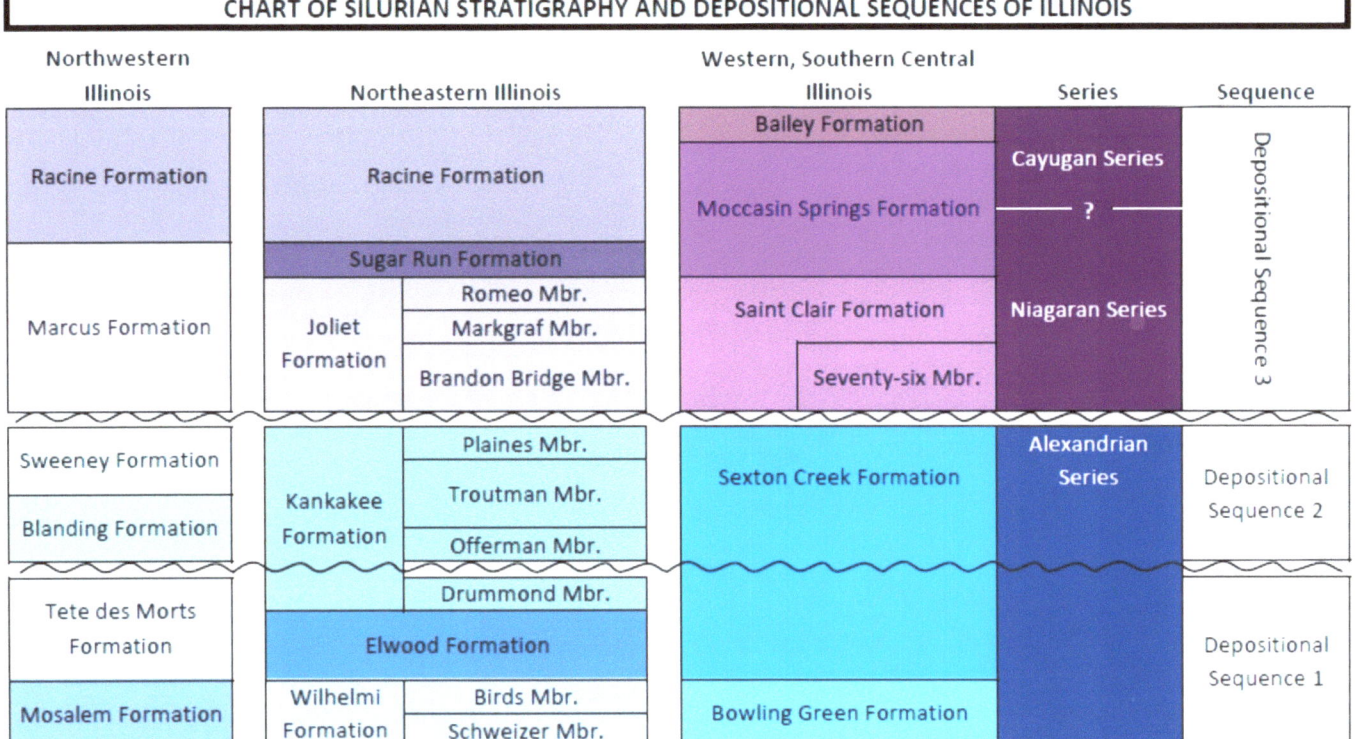

Mbr. = Member ～～～ = Unconformity

Deposition of the Bailey Formation continues up into the Devonian.
Series and Sequences apply to all regions of Illinois.
The unconformities between Sequences cross some formations, because formations are units of rock idependent of time.
Units are not representative of thickness. They are set to their equivalent units in different areas of Illinois.
Formations are selected based on the physical characteristics of a rock unit and are independent of time.
Series are "time and rock" units, usually designated based on identifiable characteristics of a rock, within a timeframe.
A sequence is bounded on top and bottom by unconformities on a regional scale.

The Silurian is also unique in its distribution in Illinois. Most formations in the Paleozoic thicken from Wisconsin to Southern Illinois. The Silurian does not. Its thickness is relatively uniform throughout the State (where it hasn't been eroded or filled Ordovician valleys). This tells us that the Illinois basin was extremely quiet at the time. For the entire Silurian Illinois appears to have been below sea level. There appears to be nothing going on tectonically in Illinois. Even the recently formed LaSalle Anticlinorium and Kankakee Arch seem to have stopped developing, at least for a while.

This isn't to say that nothing was going on. The Silurian was the time of great reefs in Illinois similar to the modern Great Barrier Reef. There are dozens of Silurian reefs throughout Illinois. They are best exposed in the Chicagoland area where they are extensively quarried. Perhaps the most famous of these quarried reefs is Thornton's Quarry, on I-80 just west of the Indiana border (Volume II, p.31-32).

The Silurian seas had much of the same fauna as did the Cambrian and Ordovician. There were two major advancements of life during this time. Modern looking fish with jaws appear and became abundant. Plants also moved to the land at the end of the Silurian but remained relatively limited until the end of the Mississippian Period.

At the end of the Silurian the seas would retreat from Northern Illinois but deposition would continue in Southern Illinois from the end of the Silurian through the Devonian. In most of northern Illinois, the Silurian-Devonian contact marked by an unconformity. In Southern Illinois, deposition was continuous and some geologic units, like the Bailey Formation, were deposited during the Silurian and the Devonian.

Location: Display case at the Illinois State Geological Survey. The unconformity between the Silurian and Devonian is often very pronounced. The light colored rock on the top is the younger Devonian and the reddish rock on the bottom is the older Silurian. The unconformity is the thin dark line that runs across the middle of the polished core.

Silurian-Devonian contact
Bartelso East Reef
Silurian
Moccasin Springs Formation

Clinton County
Depth 2559 ft

The Middle Devonian marks the end of the Tippecanoe II Sequence and the beginning of the Kaskaskia Sequence (p.14).

Outcrops of the Devonian are somewhat rare in Illinois, although it is extensive in the subsurface. Devonian rocks tend to be deeply buried by Mississippian, Pennsylvanian, and Quaternary rocks. In Northern Illinois, the rocks of the Devonian are mostly impure sandy and argillaceous carbonates, with minor evaporates, such as gypsum. The massive reefs of the Silurian were no longer being deposited but little "mini-reefs" or "reef mats" several feet high are common place.

Unlike the Silurian, the Devonian thickens greatly as you head towards Southern Illinois. The reason for the vast changes in the thickness of the Devonian is due to several low lying but very broad structural features. The high areas of land during this time are the Kankakee Arch to the north and the Sangamon Arch which trends roughly east-west in the middle of Illinois.

Corals, bryozoans
Devonian
Lingle Formation

Montgomery County
Depth 2375 ft

Location: Display case at the Illinois State Geological Survey. This polished rock core was obtained from the southern edge of the Sangamon Arch in Central Illinois. It has both corals and crinoids within. Although not from a reef, marine invertebrates still tended to live in clusters on the open sea floor. During the Devonian, reefs in Illinois had all but disappeared.

At the top of the Devonian and at the very base of the Mississippian, lie the New Albany Shale in Western, Central, and Southern Illinois. The New Albany Shale is as much as 460 feet thick. The Grassy Creek Shale of the New Albany is one of the continuous units of the New Albany, stretching from the Sangamon arch into Southern Illinois. In Southern Illinois the Grassy Creek Shale contains as much as 4% to 15% hydrocarbons and is a potential source of natural gas and oil shale. It has also served as a target for hydraulic fracturing (fracking) since the late 1950's.

The early part of the Devonian would also mark the appearance of the first land vertebrates, amphibians. Some of the amphibians would be small like today's frogs and toads. By the Pennsylvanian some would be predators over six feet long! At the very end of the Devonian, the first reptiles evolved.

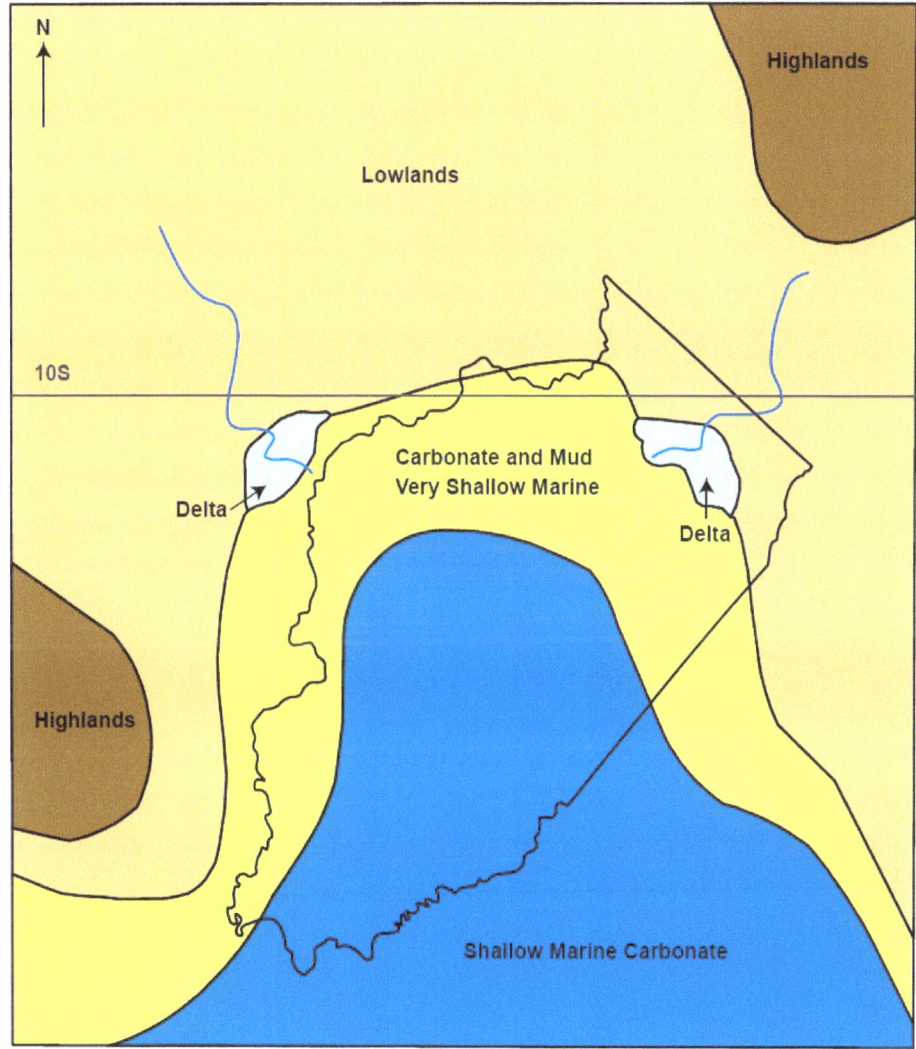

PALEOGEOGRAPHY DURING THE DEVONIAN-MISSISSIPPIAN: *This map depicts the position of Illinois in the Late Devonian through the Early Mississippian Periods. Illinois was still south of the equator and was mostly covered by a shallow inland sea. The entire State was below or near sea level. Any minor rise or fall in the sea would drown or dry out Illinois.*

Mississippian (359 to 323 million years ago), Pennsylvanian (323 to 299 million years ago), and Permian (299 to 252 million Years ago) Periods: Time of Shallow Marine Pools, Coal Swamps, and Volcanic Intrusions

The end of the Devonian and the beginning of the Mississippian marks a continuous yet pronounced change in depositional environments. The Devonian rocks are mostly marine basin deposits as where the Mississippian rocks are mostly near shore delta and shallow marine platform deposits The delta deposits are represented by the 650 foot thick Borden Siltstone. The Borden Siltstone delta is roughly horseshoe shaped and extended from the southeast in Kentucky and extended into western Illinois, terminating at the Sangamon Arch. The Borden Siltstone is surrounded and capped by the thick carbonate rocks of the Mammoth Cave Group, which are up to 2,500 feet thick in Southern Illinois!

There is an unconformity on top of the Mammoth Cave Group. It probably only represents a couple of million years of missing time, but above the unconformity the depositional setting changed once again. The rest of the Mississippian Period would see great rises and drops in sea level. This rapid fluctuation of the sea is preserved in the rocks. Most of the rocks above the unconformity are almost all near shore or terrestrial (or land) deposits. Overall the rocks of the Late Mississippian are shale and limestone, but these are cut by thick channelized deposits of sandstones. These sandstones were deposited by rivers whenever the sea level would fall. If sea level stayed low enough for a while, a paleosol would form. Mississippian paleosols are identified by their distinctive red color and root structures from the newly evolved and still small, land plants. At present, there are ten such paleosols identified in the Late Mississippian rocks of Illinois.

The upper half of the Mississippian rocks in Illinois also represents an unusual occurrence in geology. They show twenty rhythmic sea level rises (transgressions) and falls (regressions). These rocks all belong to the Chesterian Series and are extensively exposed in Southern Illinois. They are numbered one through twenty (oldest is number one and the youngest is number 20) and each represents about 850,000 years of deposition followed by 50,000 years of erosion. These cycles were probably caused by the waxing and waning of the ice cap at the South Pole. Even with this somewhat regular rise and fall of the seas, the thickness of formations varies greatly, mainly due to the depth of erosion at the end of each cycle. During the Mississippian, plants had just got a firm hold on land. In Illinois, they wouldn't begin to form large forests and swamps until the Pennsylvanian.

At the end of the Mississippian erosion would once again resume in Northern Illinois and any Late Mississippian rocks that were deposited north of the Sangamon Arch would be completely eroded. This unconformity was probably due to the reactivation of the LaSalle Anticlinorium. This reactivation would continue through the end of the Paleozoic. In Southern Illinois the forces of erosion were much weaker and in the deepest part of the Illinois Basin, deposition into the Pennsylvanian may have been continuous. The end of the Mississippian would also mark the end of the Kaskaskia Sequence in North America and the Pennsylvanian would mark the beginning of the Absaroka Sequence.

The Pennsylvanian marks the beginning of deposits that would become one of Illinois' most valuable resources, coal. The Pennsylvanian saw an explosion of plant evolution on land. Massive primitive forests covered the entire State at one point or another. These forests thrived on the shores of ancient deltas, similar to environments in the Southern United States today. Throughout most of the Pennsylvanian a vast shifting delta complex moved from Northwestern into Southeastern Illinois. Land was towards the east-northeast and the open sea was to the northwest and south. Swamps and bayous were common place on the landscape, along with large insects.

It is also during this time that a strange creature would appear in the swaps, the Tully Monster. The "Tully Monster" or *Tullimonstrum gregarium* was an unusual swimming predator with a long body and protruding eyestalks and mouth. Rarely over six inches long, the Tully Monster was probably one of the top predators of its day. It resembles no known phylum, living or extinct. Well, it didn't, until detailed studies were made of its anatomy. It turns out it had gills and chemicals in the eyes (pigment granules called melanosomes) resembling vertebrates, making it a fish. The study was released in March 2016. However, in 2019, it has subsequently been put back in the bin of the unknown, because some invertebrates (like squid and octopus) are now known to have similar chemicals in their eyes. So what was it? Who knows?

It is so rare and unique to Illinois, that it has been officially designated as Illinois' state fossil. It is found encased in the iron carbonate concretions (siderite) of the Francis Creek Shale of the lower Carbondale Formation. It is one of the unique fossils of the Mazon Creek fauna, which also include jellyfish, amphibians, fish, worms, and spiders.

PENNSYLVANIAN PALEOGEOGRAPHY: *During the Early Pennsylvanian Period, Illinois was at the equator. Glaciers covered the South Pole. As they would grow and shrink, sea level would rise and fall by about a hundred feet. This is far more than the smaller transgressions-regressions of the Mississippian Period. Since Illinois was located near the equator and below or at sea level, massive peat deposits formed in swamps as land plants got a firm hold on the land for the first time. These peat deposits would become the coals of today.*

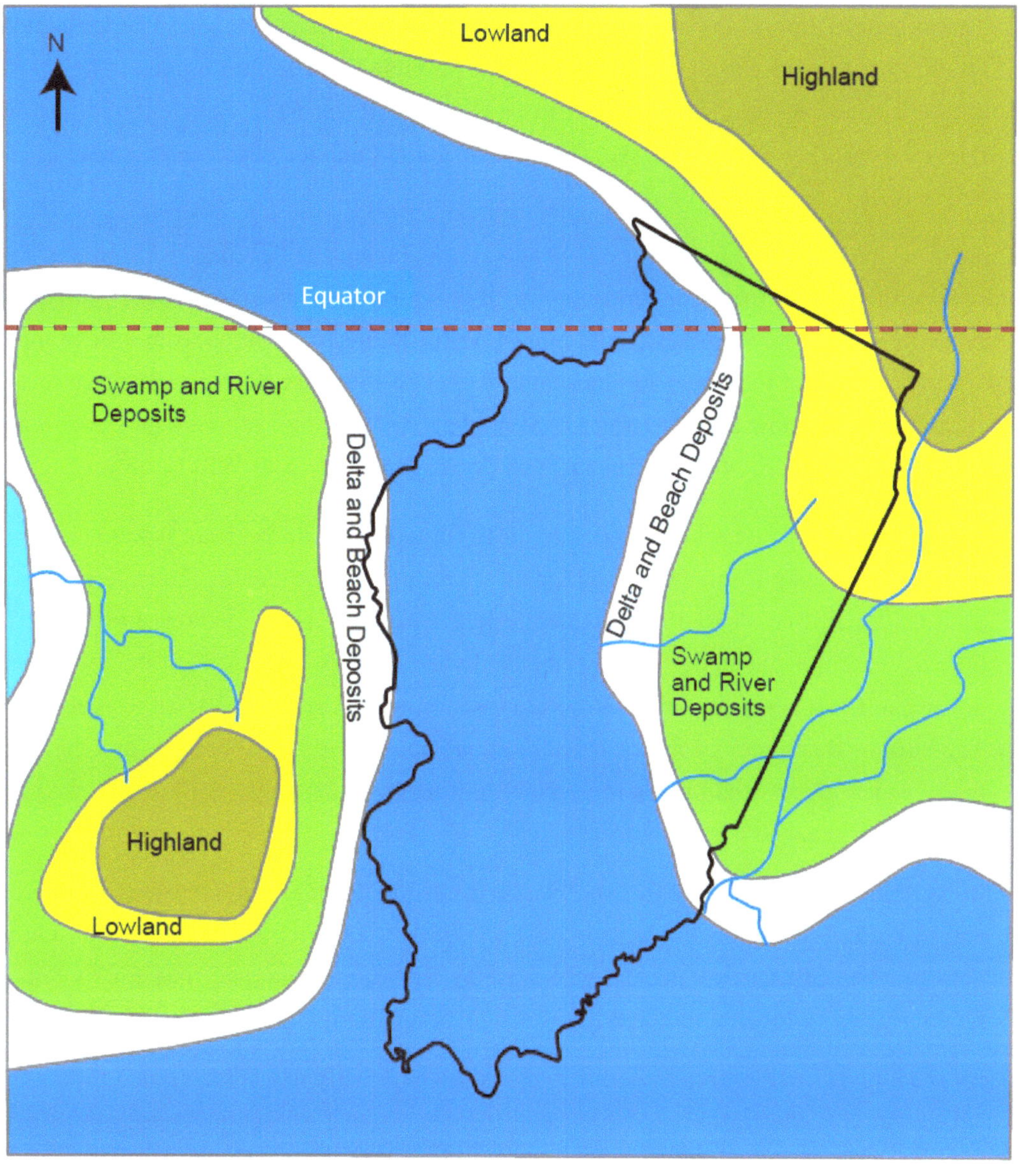

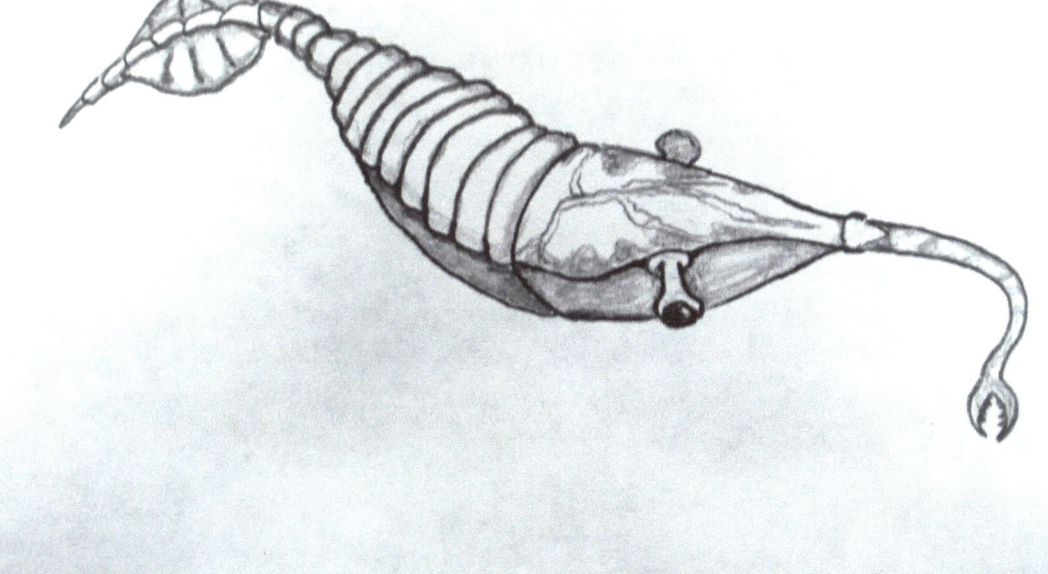

Author's (Steven D.J. Baumann) reconstruction of Tullimonstrum gregarium (Tully Monster). The mouth is the claw like feature at the end of the head. The eyes extend outward perpendicular to the head.

Tectonically Illinois was moderately active during the Pennsylvanian. The LaSalle Anticlinorium was reactivated, although not to the extent that it was in the Ordovician and Mississippian. The Mississippian Arch began to form in the area of the modern Mississippi River. The Plum River Fault Zone entered into Illinois from Iowa. Also during the end of the Pennsylvanian, the supercontinent of Pangaea began to assemble as most of the world's landmasses began to collide near the equator.

What happened between after the Pennsylvanian, during the Permian in Illinois is somewhat unclear, since no sedimentary rocks from this time exist in Illinois. Permian sediments were probably deposited in Illinois. The main line of evidence for Permian sediments is that the Pennsylvanian coals had to be much more deeply buried than they are today in order to become bituminous. Although there are no Pennsylvanian rocks throughout much of Northern Illinois, we do know that they were deposited as well. Pennsylvanian deposits commonly fill fractures and caves in the Silurian and Ordovician carbonates of Cook, Kankakee, and Will Counties.

The only record of what was happening in Illinois during the Permian is known from Southern Illinois at and surrounding Hicks Dome (Volume II, p.84). During the Permian as the supercontinent Pangaea came into being, Southeastern Illinois experienced an episode of volcanic activity centered on Hicks Dome in the Shawnee National Forest. This igneous activity was associated with the New Madrid Rift and led to heavy faulting in the area. Hicks dome itself is a crypto-volcanic structure that formed when magma tried to reach the surface but was unsuccessful. The bulk of the igneous activity at Hicks Dome appears to have occurred at a depth of 11,000 feet or more. Only a few igneous dikes would reach the surface and most of them are vent breccias that formed when pressurized gases, not magma, reached the surface.

In Northern Illinois, the Kankakee Arch continued to form as the Sandwich Fault Zone began offset rocks parallel to and just south of the Arch. Although the age of the Sandwich Fault Zone is not known for sure because no rocks younger than Silurian are present within it, the Permian is a logical time for its initial formation coinciding with the igneous activity to the south and the closing of the Illinois Basin.

By the end of the Permian Period Illinois would be mostly dry land. It also marks the end of the Paleozoic Era worldwide and the Absaroka Sequence in Illinois. A sort of second "great unconformity" began to form, since no new rocks would be deposited until the Cretaceous.

During the Permian mammal like reptiles (proto-mammals) such as *Dimetrodon* evolved. These new animals would give rise to modern day mammals but they would not survive the Permian. Life would also undergo another dramatic change. The end of the Paleozoic Era saw the worst mass extinction in the prehistory of the Earth. Over 90% of sea and land animals became extinct. This was a far worse than the asteroid impact that greatly contributed to the extinction of the non-avian Dinosaurs. Ironically, the Dinosaurs were wiped out by a mass extinction, yet it is the mass extinction at the end of the Paleozoic that opened a window for them to dominate the land during the Mesozoic. In southern Illinois at least two domes formed from igneous activity deep under the surface. The associated igneous activity helped deposit our State mineral, fluorite. Fluorite has been mined from southern Illinois. Some beautiful specimens have been pulled from the mines. The mining of fluorite essentially stopped in the 1950's, but recent demand has perked the interest of mining companies once again.

Location: Open display in the Illinois State Geological Survey Annex. This fluorite specimen was obtained from closed mines in fluorspar district. It shows perfect purple crystals of fluorite. Fluorite can be almost any color. In Illinois, purple and deep blue are the most common colors.

Phanerozoic Eon, Mesozoic Era: Land Animals Reach Their Apex

Triassic (252 to 201 million years ago) and Jurassic (201 to 145 million years ago) Periods: The Rocks that Never Were

At the beginning of the Mesozoic, during the Triassic Period, Illinois was high above sea level, as it is today. Illinois was landlocked for thousands of miles as the supercontinent of Pangaea surrounded it.

During the Triassic, Illinois was probably as high above sea level as it is today, but it sat closer to the equator. The environment supported desert and savanna fauna, which was a stark contrast to the coal swamps of the Pennsylvanian. The Triassic was an active time of erosion in Illinois. Any Permian sedimentary deposits were completely eroded along with the Pennsylvanian in the extreme northern part of the State. It was during this time that the Pascola Arch fully formed in Missouri and Tennessee, as Pangea began to breakup, permanently closing off the Illinois Basin from the open seas...at least for the most part.

Between the Triassic and Jurassic, Pangaea began to split apart. This created new ocean basins and younger, more buoyant ocean crust. The new ocean crust allowed the seas to cover much of North America again. Alas, Illinois was still high and dry, but lower in elevation relative to sea level. The Dinosaurs roamed Illinois, but there are no know fossils of them anywhere in Illinois. The dominant process of erosion was not conducive to the preservation of life during the Mesozoic in Illinois.

Cretaceous Period (145 to 66 million years ago): Illinois was a State with Ocean Front Property

The supercontinent of Pangaea continued to split apart the shallow seas covered more and more of North America. By the end of the Cretaceous, The seas had once again entered Illinois. This worldwide rise in sea level is known as the Zuni Sequence. This new rise in sea level allowed for the Western Interior Seaway to enter the very western tip of Illinois during brief periods of time. This allowed for rivers to deposit sand, silt, and gravel in Western Illinois. These deposits are still preserved as the Baylis Formation in Adams and Brown Counties (Volume II, p.60). To the south would become the Gulf of Mexico, reached all the way up to Illinois from the south as the Mississippian Embayment formed.

PALEOGEOGRAPHY DURING THE LATE CRETACEOUS:
Illinois during the Late Cretaceous Period about 90 million years ago. The Greenhorn Seaway may have reached the western tip of Illinois. It usually was much further west, near the western border of Iowa.

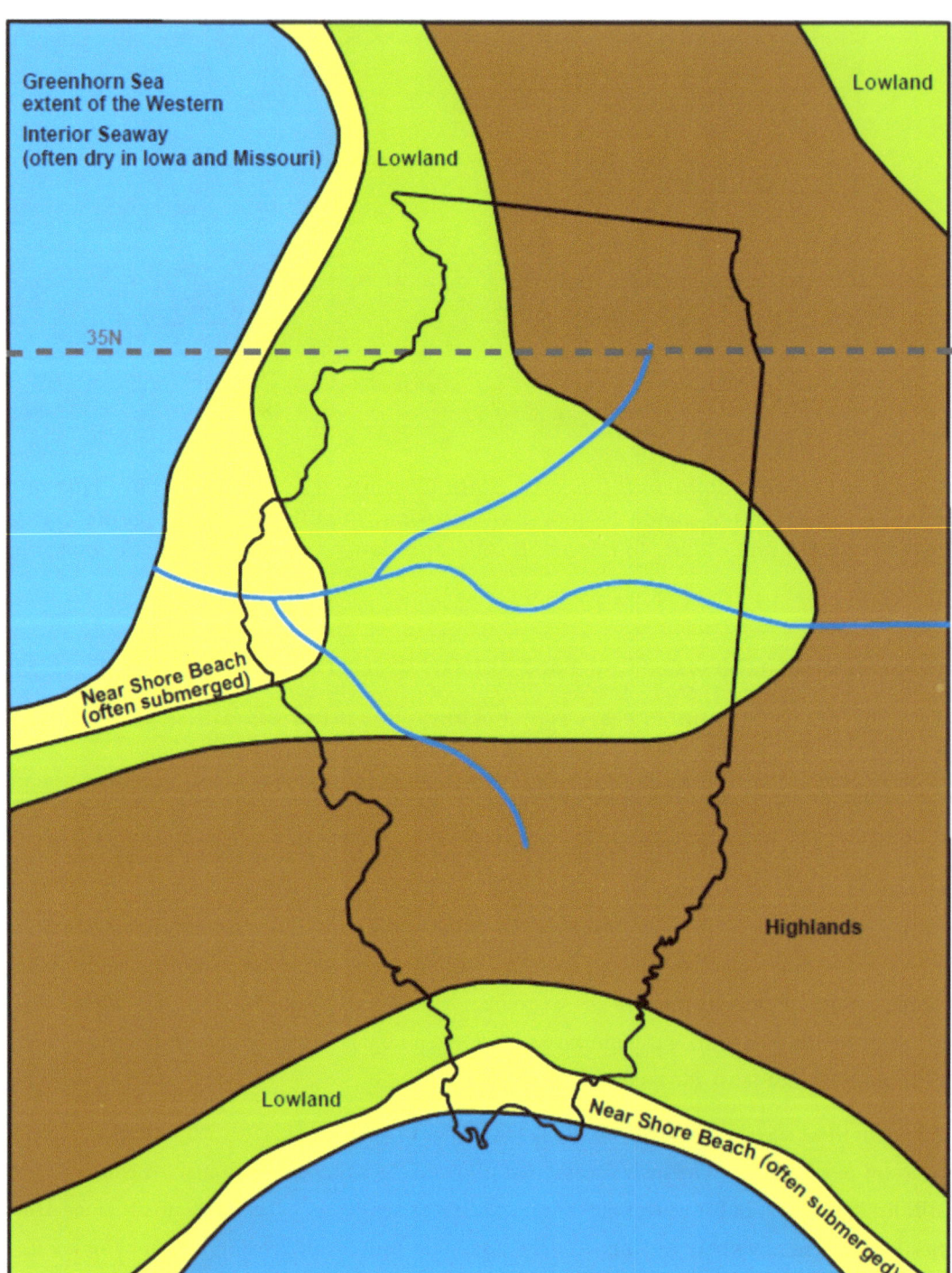

In Southern Illinois, the beach sands of the McNairy Formation and the river gravels of the Post Creek Formation (which likely correlate with the Baylis Formation in Western Illinois) were deposited. Although we know that dinosaurs roamed Illinois during this time, their fossils have yet to be found. The closest known dinosaur fossils are from Marble Hill in Southeast Missouri, about 30 miles west of Alexander County. Beach and river deposits do not preserve large fossils very well, unless they are quickly buried. The Cretaceous deposits in Illinois were never significantly buried. We know this because they are not lithified. In other words the Cretaceous sediments were never turned into hard rock.

Other than the formation of the Mississippi Embayment, the only other significant tectonic event to occur was in Southern Illinois. The Precambrian Reelfoot Rift became reactivated, and it remains active today, generating earthquakes.

At the end of the Cretaceous, and into the Paleogene, new ocean crust formed caused by the break-up of Pangaea and the ocean basins deepened, causing sea level to drop worldwide. The Zuni seas would have one more major transgression before leaving North America for good.

Phanerozoic Eon, Cenozoic Era: A New Dynasty Dawns

Paleogene and Neogene Periods (66 to 2.6 million years ago): Illinois Begins to Take on a Familiar Appearance

At the very beginning of the Cenozoic the Zuni seas deepened from Southern Illinois and may have reached as far north as Marion County. This last transgression of the oceans into Illinois would deposit the Midway Group. The Midway Group consists of the Clayton and Tuscaloosa (now called the Porters Creek) Formations, which stretch more or less continuously from Southern Illinois into the Coastal Plain of Georgia. The Clayton is definitely marine in origin and contains dark green glauconite, snails, lobsters, sharks, fish, and turtle fossils.

About 55 million years ago the Zuni seas would permanently disappear from Illinois, leaving Illinois high and dry right up to the present day. There are later non-glacial Cenozoic deposits in Illinois such as the Mounds Gravel, Grover Gravel, Wilcox Formation, and Claiborne Formation. These are mostly river and floodplain deposits. The Mounds Gravel was deposited by the Ancient Mississippi and Ancient Iowa Rivers. These sediments may actually have been deposited right up to the beginning of the Quaternary.

No new geologic structures were formed during this time, and Illinois began to take on a modern appearance. Mammals and birds began to populate Illinois and forests began to spread throughout the State. Primitive horses and even rhinos roamed Illinois along with deer and large cats. Flowers, which evolved during the Cretaceous, became plentiful. Primitive grasses began to evolve as prairies began to cover small parts of Illinois during the latter half of the Cenozoic. During the Paleogene, Illinois was warmer than today. The climate was much more similar to that of Alabama and Mississippi. Winters were short with little snowfall.

CHART OF PHANEROZOIC GEOLOGIC UNITS IN ILLINOIS

ERA	PERIOD	MILLIONS OF YEARS AGO	GEOLOGIC UNITS DEPOSITED	FOSSILS
CENOZOIC	QUATERNARY	2.6	See Quaternary Chart	
CENOZOIC	PALEOGENE + NEOGENE		Mounds-Grovel Gravel Formations Wicox Formation Porters Creek Clay Clayton Formation	Shark teeth, lobsters, turtles
MESOZOIC	CRETACEOUS	66.0	Owl Creek Formation McNairy Formation Baylis Formation	No macro-fossils
MESOZOIC	JURASSIC	145 201	No known deposits	
MESOZOIC	TRIASSIC	252	No known deposits	
PALEOZOIC	PERMIAN	299	Igneous Intrusions	No fossils
PALEOZOIC	CARBONIFEROUS (PENNSYLVANIAN)		Patoka Formation (Trivoli Sandstone Member) Shelburn Formation (Copperas Sandstone) Carbondale Formation (Vermilionville Sandstone) Carbondale Formation (Francis Creek Shale) Carbondale Formation (Colchester Coal) Tradewater Formation (Seville Limestone) Tradewater Formation (Bernadotte Sandstone) Tradewater Formation (Pope Creek Coal) Tradewater Formation (Babylon Sandstone) Caseyville Formation (Pounds Sandstone) Caseyville Formation (Battery Rock Sandstone)	burrows plankton plants "Tully Monster" crinoids, shark teeth, braciopods, bryozoans, plants
PALEOZOIC	CARBONIFEROUS (MISSISSIPPIAN)	323	Kinkaid Limestone Degonia Sandstone Palestine Formation Menard Limestone Goiconda Formation West Baden Formation (Ridenhower and Bethel) Paoli Formation Aux Vases Sandstone Ste. Genevieve St. Louis Limestone Salem Formation Warsaw Formation Burlington-Keokuk Limestone	crinoids, brachiopods, sponges, gastropods, Archimedes
PALEOZOIC	DEVONIAN	359 419	St. Laurent Formation (a.k.a. Alto and Lingle) Grand Tower Limestone Bailey Formation	same as Silurian
PALEOZOIC	SILURIAN	444	Racine Formation / Moccasin Springs Formation Sugar Run Formation / Moccasin Springs Formation Joliet Formation / Saint Clair Formation Kankakee Formation / Sexton Creek Formation	trilobites, coral, brachiopods
PALEOZOIC	ORDOVICIAN	485	Maquoketa Group Galena-Platteville Groups / Kimmswick Limestone Glenwood Formation Saint Peter Sandstone Shakopee Formation New Richmond Formation	cephalopods, stromatolites, trilobites, coral
PALEOZOIC	CAMBRIAN	542	Potosi Formation Franconia Formation Ironton-Galesville (subsurface only) Eau Claire Formation (subsurface only) Mount Simon Formation (subsurface only)	fossils are rare mostly marine burrows, some trilobites
	PRECAMBRIAN		See Precambrian Chart	None

Geologic units listed above are by no means all of the geologic units in Illinois. Only the major geologic units mentioned in the text are listed above.

Then a change occurred far from Illinois that would affect the entire globe. Around 20 million years ago a great freeze began. Antarctica began to freeze over year round and slowly accumulated vast ice sheets. Illinois reached a latitude of about 40° north at the Northern State line. The present northern border of Illinois is at 42.5° latitude. This would help set the stage for the vast ice sheets that would cover Illinois during the Quaternary.

Quaternary (2.6 million years ago to today) Period: Age of Ice and Men

The Quaternary was a time of vast change not only for Illinois but for the world. It is a time when vast continental glaciers would expand from the northern polar region and reach as far south as Southern Illinois, to what is now known as the Shawnee National Forest. These glaciers would grow larger and then retreat at different times. In the northern hemisphere, the Quaternary is known for the Ice Age.

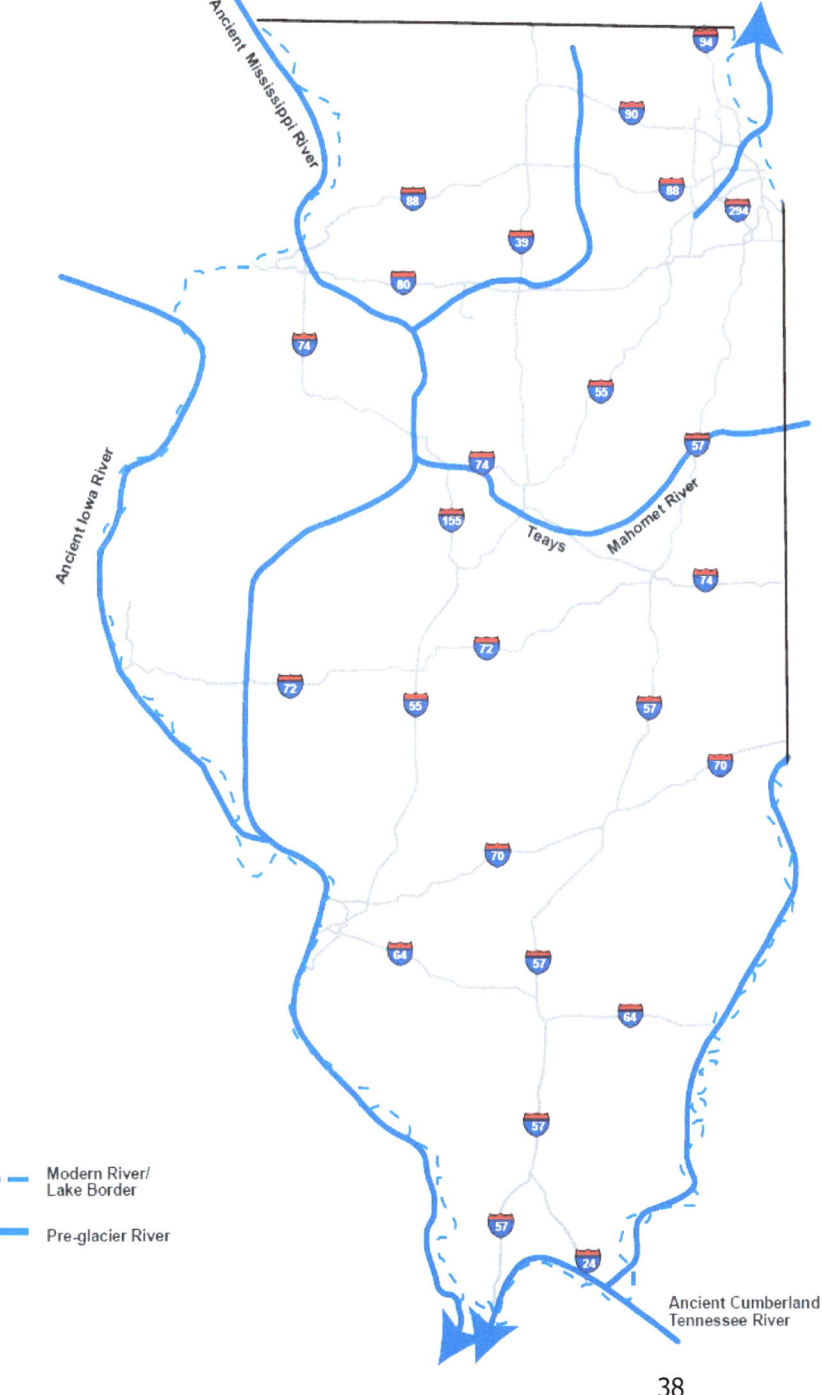

Two million years ago, Illinois was essentially at its present location. It was a very different landscape. There were deep valleys and high ridges; there was little of the vast flat areas that we see today. Also, the major waterways in Illinois looked very different than they do today. Lake Michigan was a river system. There was no Illinois River. The Mississippi River flowed further east, occupying part of the Illinois River's modern path. There was no Ohio River as we would recognize it today. The now buried Teays-Mahomet River dominated Central Illinois.

It can be hard to visualize 300 to 1,000 foot high glaciers covering Illinois. We tend to think of them in the cold places at the planet's poles or high in the mountains. Illinois is covered in deposits that resemble those left by massive glaciers, so they had to be here. What could have caused them? There are several theories on what causes glaciers to expand and retreat. They include minor changes in the Earth's orbit, the concentration of continents in one hemisphere or the other, volcanic activity, and the presence or lack of continental seaways. All of these things likely contribute to the advance of glaciers.

The glaciers would extend into Illinois during a minimum of three times during the Quaternary, and likely many more. These major ice advances occurred during the Pre-Illinois, Illinois, and Wisconsin Episodes. Although glacial advances probably took place in Illinois during the early Pleistocene (the first part of the Quaternary), we only have direct evidence of their presence spanning back to about 850,000 years ago.

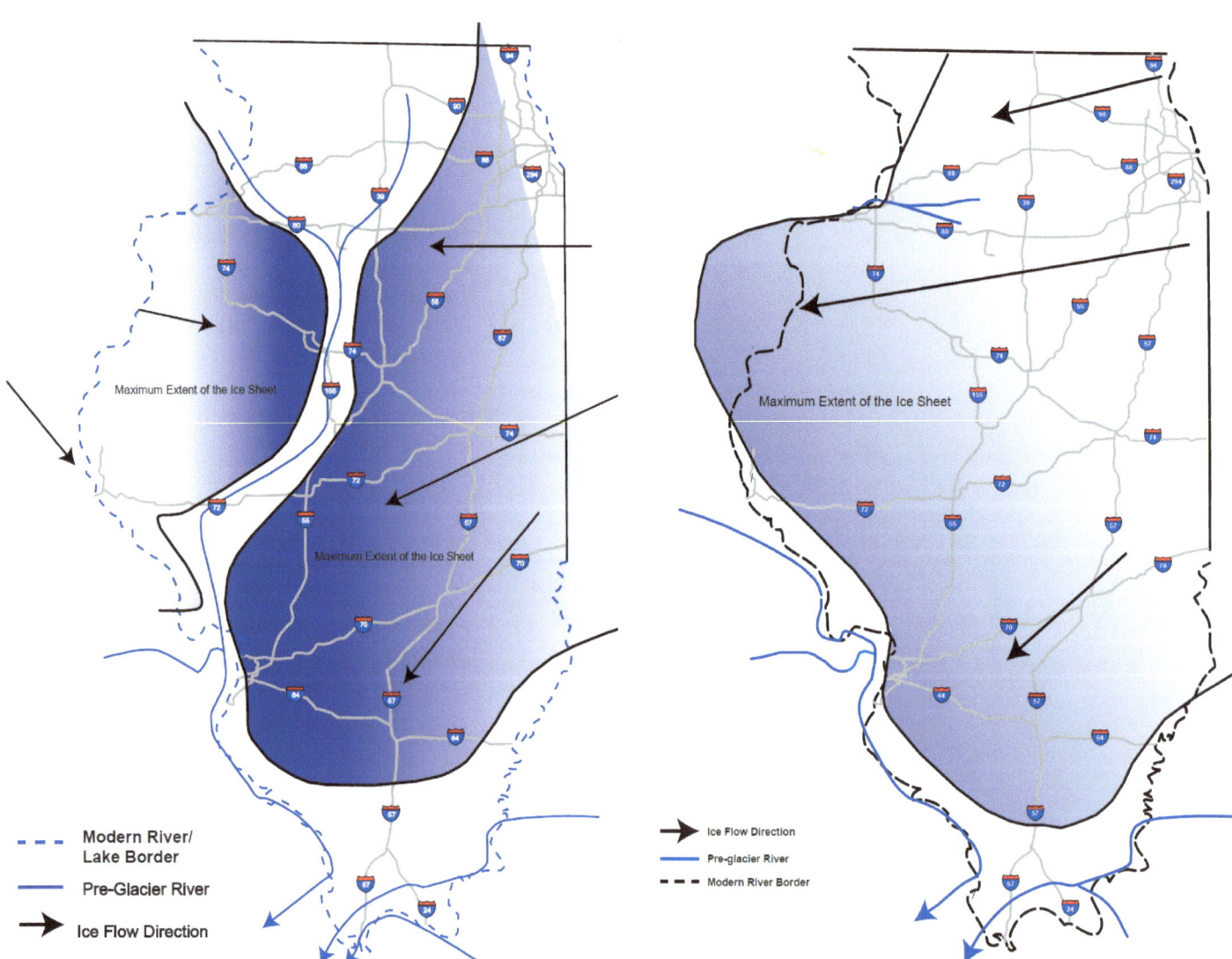

The first preserved evidence of ice covering Illinois occurred about 850,000 years ago, during the Pre-Illinois Episode. There were two major glacial lobes. One came from the west-northwest and the other from the east-northeast. When the ice receded 400,000 years ago, the Teays-Mahomet River would be forever buried.

About 150,000 years ago the glaciers returned to Illinois during the Illinois Episode. This time the ice descended from only the east-northeast and reached their maximum extent of coverage. Even though the ice covered the Ancient Mississippi River, when it retreated, the river would resume its pre-glacial path.

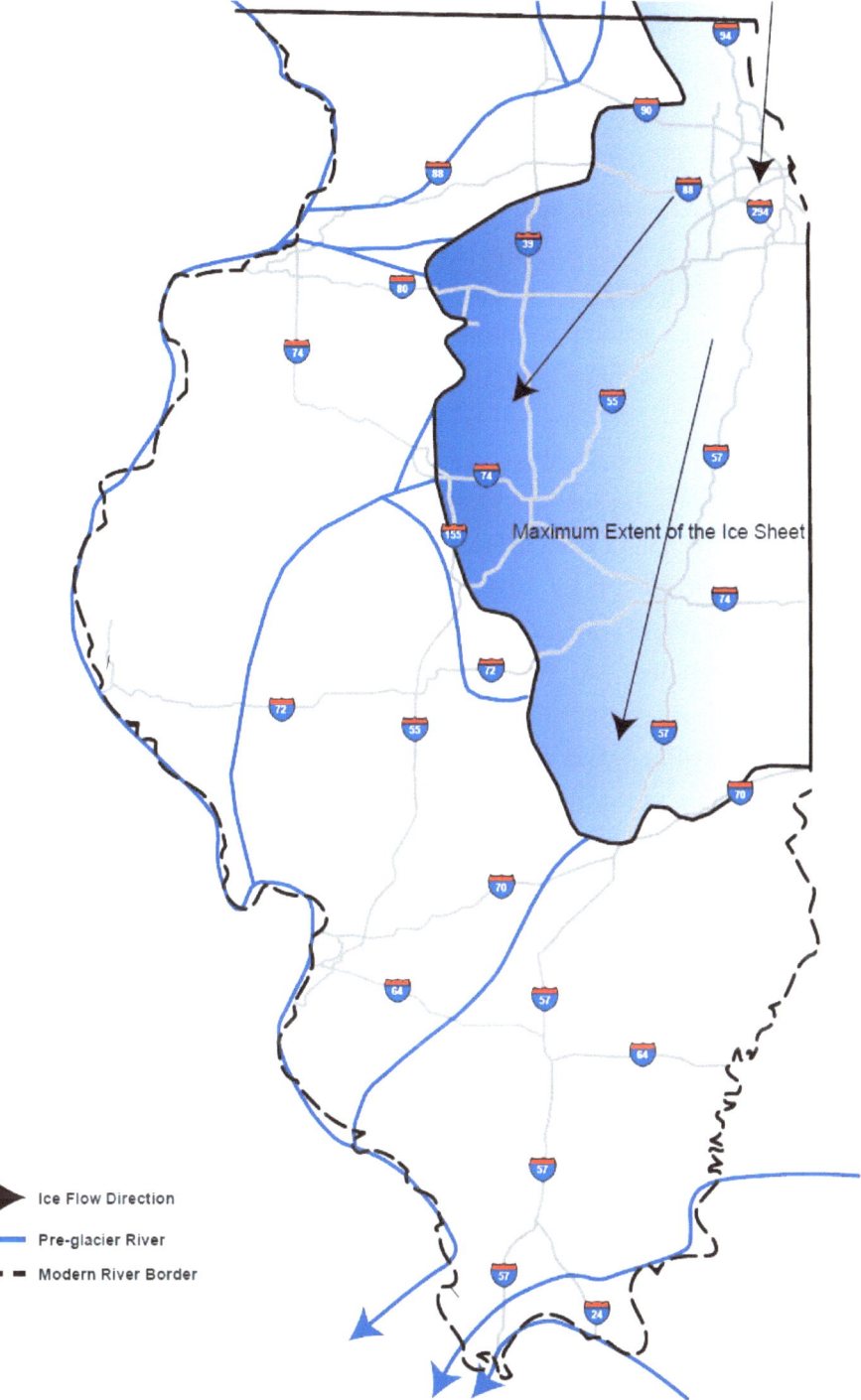

25,000 years ago the glaciers would readvance into Illinois. Although they would not cover as much area, they had a far greater impact on the landscape. After the ice melted, Lake Michigan was left behind. The Mississippi River was pushed westward to its present position. The modern Illinois River took shape. The modern Ohio River still flowed through Illinois, but it was north of its present position through the modern Cache Valley. It wouldn't take on its modern shape until long after native people settled the area.

There is a misconception that glaciers act like giant bulldozers just scooping up and shoving the surface forward. They actually erode rock by a process called plucking and abrasion. They act more like coarse grained sand paper moving over soft wood and pick up the tiny loose pieces at the surface.

Glacial deposits can be grouped into several basic types. Diamicton (the modern roughly equivalent term for drift and till) are deposits that are mostly fine grained, poorly sorted, and usually contain pebbles to boulder size rocks that was deposited directly by the glaciers. Outwash is usually coarser sediments of sand and gravel that are sorted, layered, and deposited in flooding events as ice melted. Lacustrine (lake) deposits of laminated sands, silts, and clays deposited in glacial and slack water lakes. Finally loess, which is mostly fine silt sized (rock flour) to very fine sand that has been transported by wind.

We know that there have been several advances and retreats in Illinois as opposed to a single event because of geosols. These ancient soils formed after a glacier retreated during a warm period known as an interglacial episode. There are three named interglacial episodes in Illinois. They are from youngest to oldest, the Hudson (or Modern), Sangamon, and Yarmouth Episodes. There are five identified interglacial geosols recognized at present.

In Illinois, glaciers tend to leave a generally flat landscape made up of diamicton, in what is called ground moraine. Ground moraine is what anyone driving long distances in Northern Illinois on the expressway sees when you observe vast flat farm fields, especially on I-55 and I-57. Although there are many features that stand-out against the flat plains. Ridges and hills of isolated sand and gravel often stand in contrast to the flat surrounding areas. The ridges are called eskers and form as stream deposits at the base of glaciers. Kames are round hills of sand and form in several ways. The most common way is as a small delta at the front of a glacier. As you drive the flat areas of Illinois, you will also notice that low lying, yet continuous long ridges. These are usually end moraines. They are composed of mostly diamicton and they were deposited at the leading edge of stagnant glaciers. Illinois also boasts many isolated natural lakes. These lakes are most common in the northeastern part of Illinois. These lakes are formed in one of two main ways. Some formed as kettles. Kettles form when ice that was trapped in moraines as the glaciers retreated. The trapped block of ice would later melt forming a shallow depression in which a small lake would form. Kettle lakes are very short lived and most are gone. Some still exist, such as the ones that form the Chain 'O Lakes. Others form when a stream cut by melting glaciers become naturally dammed. These are called slack water lakes. Other prominent features are formed by loess deposits, which usually form bluffs near major rivers and were deposited by wind on barren landscapes as the glaciers retreated.

There was a road cut exposed along Illinois Route 3 called the Thebes Road Cut Section. This outcrop was briefly exposed during road construction in 2006. It is a significant road cut because it had all three major geosols on top of older loess deposits in a section no more than 30 feet high. Rarely are glacial geosols preserved stacked up one on top of another. The outcrop has since been landscaped over.

Although Quaternary Ice Age deposits almost 90% of Illinois, and left sediments hundreds of feet thick, outcrops are rare. This is because they are usually covered by vegetation, form low slopes, or buried under human built structures. The best outcrops are usually fresh road cuts, quarries, and stream cut banks.

The glaciers changed the Illinois covered landscape in major ways. Prior to the Ice Age most of Illinois resembled the driftless areas of Northwestern Illinois and western Tennessee. The glaciers did more than flatten Illinois. They changed the course of the Mississippi, Ohio, and Iowa Rivers. They also forever buried large rivers such as the Mahomet River, which rivaled the Mississippi River in size. Even Lake Michigan owes its very existence to the glaciers. The Lake Michigan of today is very different from the Lake Michigan of 14,000 years ago. At times it has been significantly higher, by as much as 60 feet. At other times the Lake Michigan Basin has been almost dry. There are large deposits of sand in the Chicago Region which have either served as beach fronts or spit islands when Lake Michigan stood much higher. This period of time is called the Lake Chicago Stage.

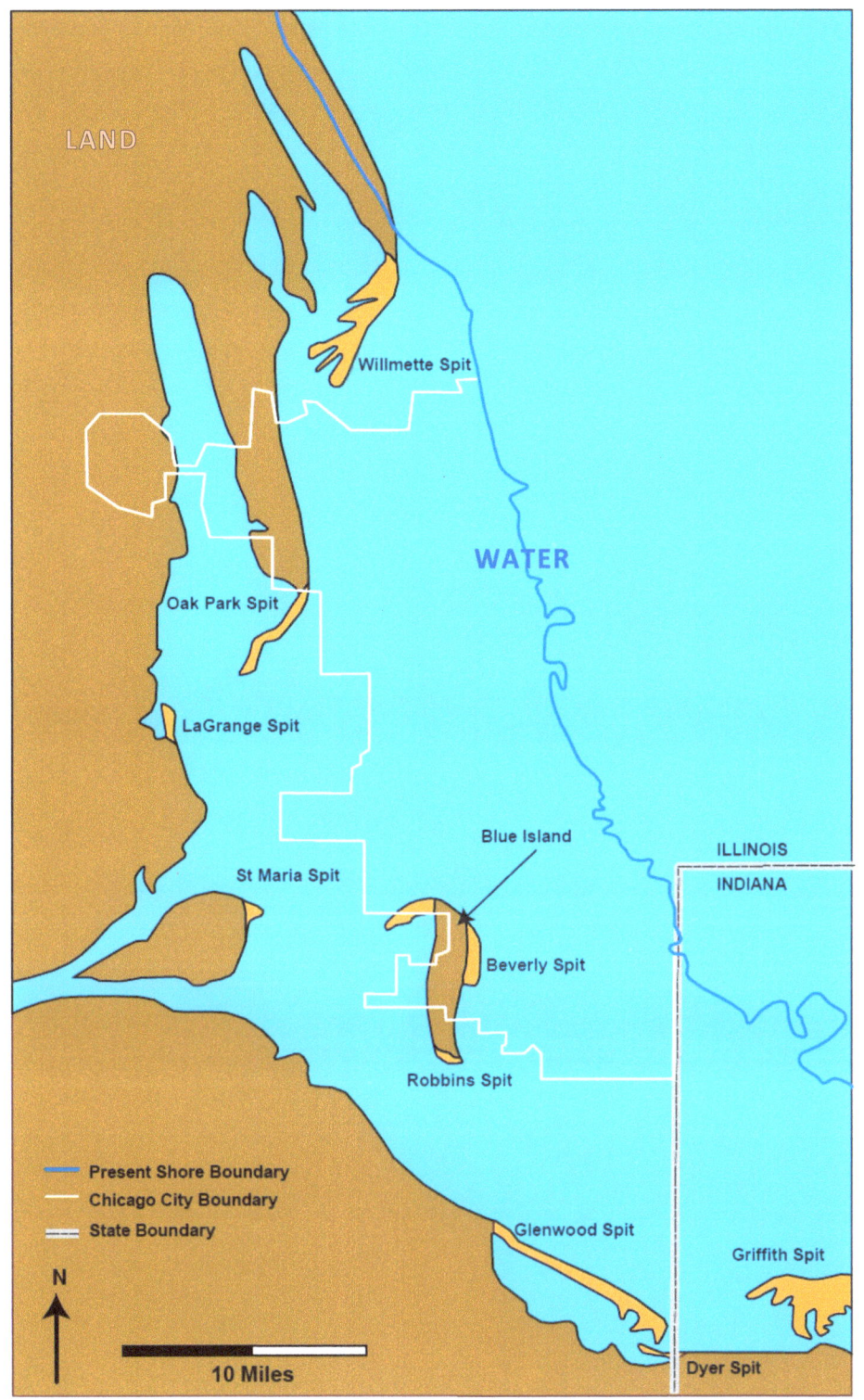

This is a map of the Chicagoland area around 14,000 years ago, during the Lake Chicago Stage. At this time, the glaciers had retreated north and were melting very fast. This caused the lake to rise significantly higher than it is today. At this time Blue Island was an actual Island. This situation would persist for about another 1,000 years.

The animals and plants during glacial advances are similar to tundra life forms today. Some such as the Caribou are still around. Others, such as the Wholly Mammoth and Saber Tooth Cats are extinct. The arrival of anatomically modern humans (about 150,000 years ago) on the planet occurred during the Quaternary and by the Holocene they would spread throughout the globe. Other than making tools and weapons, ancient humans would also become the first species to genetically engineer plants. Corn (or maize) owes its existence to the Native Americans. About 9,000 years ago, they began to selectively breed a common grass called teosinte. After generations of manipulating teosinte the first corn would appear, leading to an agricultural revolution that is still around today. Humans may be responsible for corn, but we are not responsible for everything that has happened during the Quaternary.

The vast majority of Illinois is covered by glacial deposits. They cover about 90% of the state. They not only cover the surface, but they go into the subsurface pretty deeply in places. Often the glaciers would advance and retreat, not only leaving behind end moraines, eskers, lakes, etc.; but they would also deposit sediments in the subsurface. This would often fill the old paleo-lowlands with sands and gravels (Henry Formation) if they couldn't flatten the topography. Below is a stylized block diagram through Moraine View Recreation Area, in McLean County. The park surrounds Dawson Lake. The park is mentioned in volume 2 (p. 69), but the diagram is included here, in order for you to get a sense of the subsurface glacial deposits. A combination of maps were used to generate it (p.65). It may be slightly off from what is actually in the subsurface. Most of the glacial sediments in the area were deposited between 11,500 and 25,000 years ago (p. 44). The top bedrock was deposited between 300 and 309 million years ago (p.37)

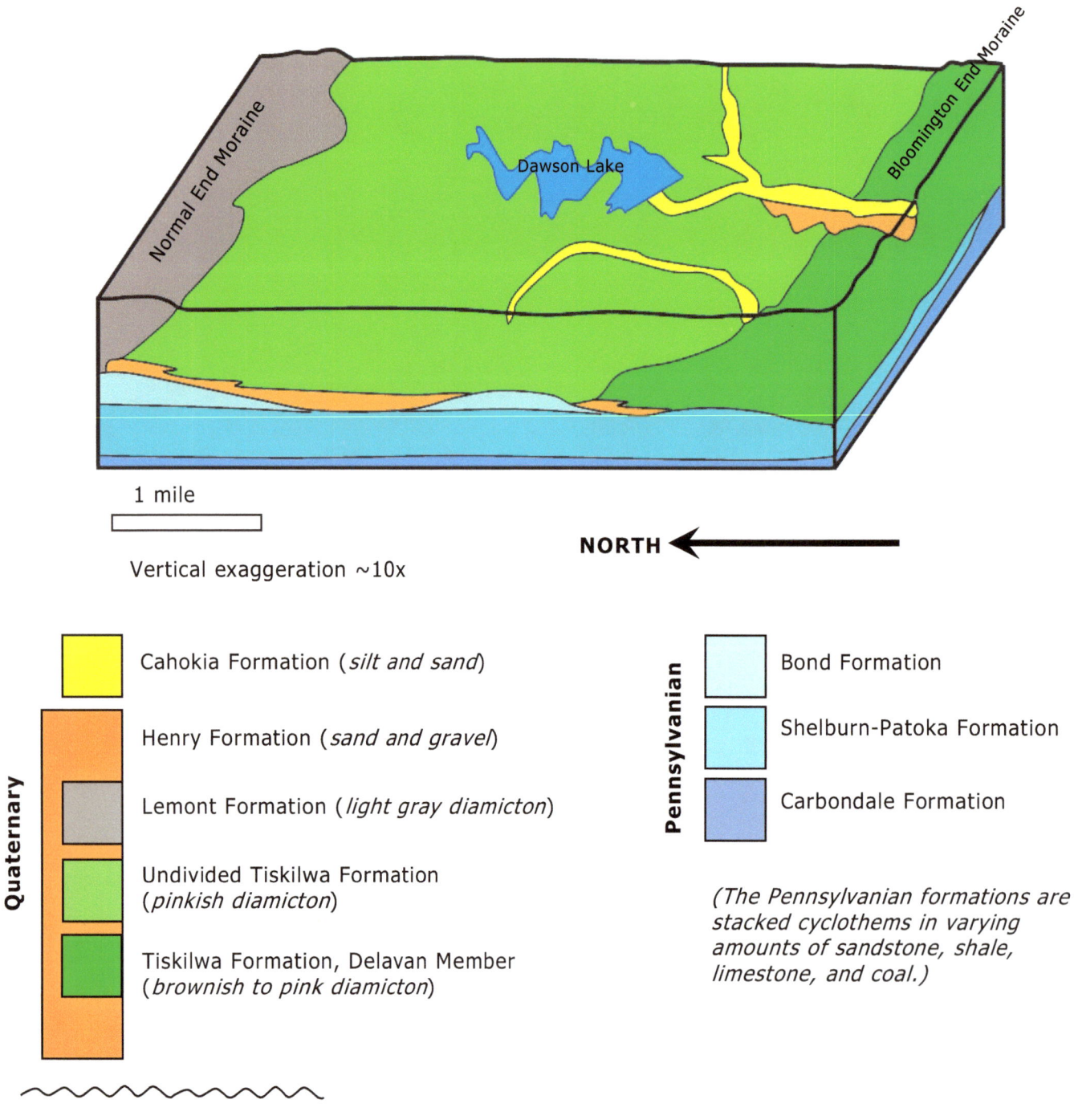

1 mile

Vertical exaggeration ~10x

NORTH ←

Quaternary
- Cahokia Formation (*silt and sand*)
- Henry Formation (*sand and gravel*)
- Lemont Formation (*light gray diamicton*)
- Undivided Tiskilwa Formation (*pinkish diamicton*)
- Tiskilwa Formation, Delavan Member (*brownish to pink diamicton*)

Pennsylvanian
- Bond Formation
- Shelburn-Patoka Formation
- Carbondale Formation

(The Pennsylvanian formations are stacked cyclothems in varying amounts of sandstone, shale, limestone, and coal.)

The Quaternary has also seen major swings in climate. There is a large misconception among people is that the climate of the past 200 years has been "the status quo". Nothing could be further from the truth. The Earth is in a state of flux and the climate of today is not the climate of the past. We are in an interglacial episode and the Ice Age is still going on. Our modern influence and conception of climate is only part of a larger picture extending back billions of years. The last interglacial period, from about 75,000 to 125,000, was warmer than it is today. We look at the polar regions and assume that they have always been covered by ice. In reality, the opposite is the case. Earth does not usually have ice caps. Until Antarctica froze over, the last time Earth had ice caps was probably during the Permian, but there was no ice age. The last time Earth experienced and actual ice age was during the Pennsylvanian. That isn't to say that humans have not contributed to the recent temperature changes. Human beings have a profound influence on the climate, but people are not necessary to facilitate climate swings. The further we go back in time the more we realize that Earth is typically a warm planet. Life is far more diverse in warmer areas than in the polar regions.

The end of the Wisconsin Episode marks the end of the Pleistocene Epoch and the beginning of the Holocene Epoch, which is the present geologic epoch. Most geologists who study ice ages think that the Holocene is just another interglacial episode.

In Illinois the Holocene began 11,500 years ago as the last continental glaciers melted from North America. The actual beginning of the Holocene varies slightly from place to place, depending on the position of global ice sheets. As the glaciers melted back, they left a barren landscape dotted with glacial and kettle lakes. With no vegetation to anchor the soil in place, vast windblown sediments would roll across the landscape before being deposited in low lying areas as dunes and loess.

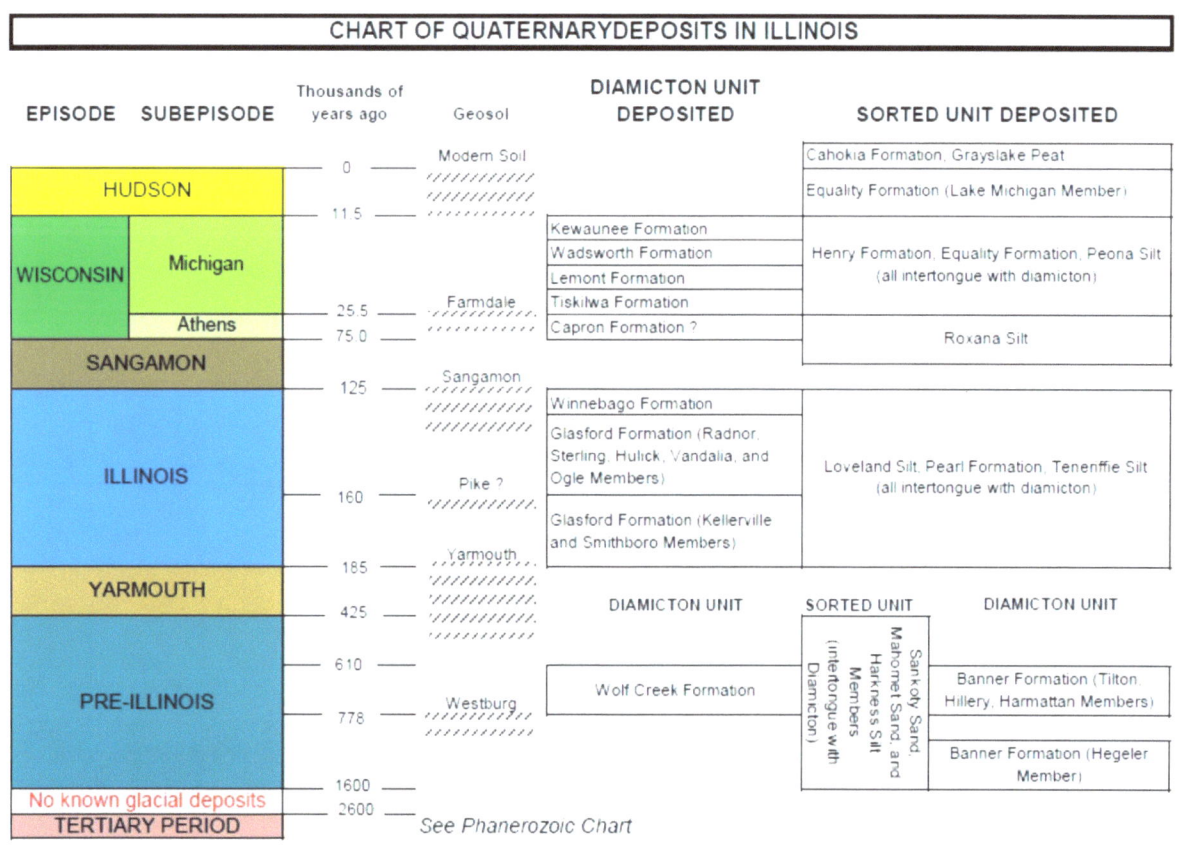

During the Wisconsin and Illinois Episodes the glaciers advanced from the northeast. During the Pre-Illinois Episode, glaciers advanced from the northeast (Banner Formation) and the northwest (Wolf Creek Formation).

Most major, well established geologic units for the Quaternary are included above. Glacial advances are blues or greens. Interglacial episodes are brown or yellow.

At present, there is only evidence for ice covering Illinois back to 1.6 million years ago. It is possible that Illinois was ice free from 1.6 to 2.6 million years ago.

GEOBIT: ERRATICS, THE ODD STONES FROM FAR AWAY

Erratics are rocks ranging from cobbles to boulders that are not local to an area. Some were carried by the glacial ice from further upstate. Others were transported by the ice as far as Ontario. We notice the large ones at the surface. Some remain buried until erosion, a farmer's plow, or backhoe unearths them. The largest erratic in Illinois is a pink granite weighing about 100 tons and resides on private property in Jefferson County. It is about 22 feet by 11 feet by 10 feet in size.

This six foot long gray granite erratic lies in the Vermilion River at Matthiessen State Park. The erratic fell from the 100 foot cliffs to the north and rolled all the way into the middle of the river. This erratic was not at its present resting place in 2012. It likely fell from the cliffs and was transported by the river during the floods that occurred throughout Northern Illinois in April 2013.

A piece of large dolostone ominously sits on the west side of Barrington Road about 1,600 feet south of the junction with IL-58. This rock is huge. It is about 12 feet long by 9 feet wide by 5 feet tall. There is another, smaller boulder exactly 200 feet north of this one. Is it an erratic or is it a boulder that moved here by people from somewhere else? The answer is both. It was originally unearthed from glacial deposits of a local farm and transported to its present location before the area was designated as forest preserve.

GEOBIT: FIELD WORK AND GEOLOGIC MAPS

One of the most difficult and important tasks that geologist do is geologic mapping. Geologic maps show the distribution of formations at and below the surface. Their main purpose is to aid in the planning of infrastructure, predicting the paths of floods, potential for environmental contamination, the exploration of economic resources, and research.

Data for generating geologic maps is obtained through a variety of techniques. One important technique that is always used is field mapping. Field geologists will take their gear and often explore every outcrop, every stream cut, and every other corner looking for rocks exposed at the surface within the area they are mapping. They will then document as much as physically possible. The location, the description of the rocks, photographs, samples, and structure of the rocks, are some of the things recorded during field work.

Accurate field notes are crucial to creating a good geologic map. In most areas of Illinois, exposures are rare, if they are present at all. Also most of Illinois' topography is due to the ice age. The bedrock was deposited in a very different environment. We cannot count on the surface to accurately reflect the subsurface geology, even if glacial deposits are thin. This means that we need to rely on other techniques. We have to rely heavily on drilling logs, seismic surveys, and mining records. As we dig deeper and deeper into the Earth to get our resources, subsurface information becomes more and more important.

Field mapping is still very important. It just means that you may be out logging drilling cores instead of hiking mountains and crawling through valleys. In the driftless areas of Illinois, walking a mapping area is still a very important part in making a geologic map.

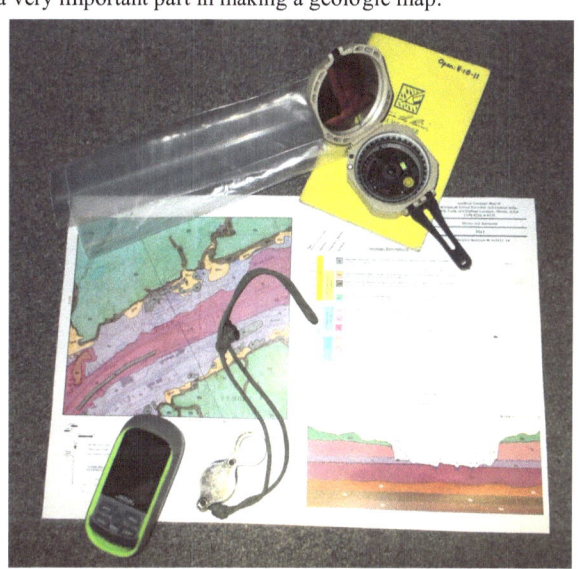

A simplified geologic map of Keepataw Forest Preserve (created by the author) under several geologic field items. The item at the top left is a sample bag. The two items on the top right are a Brunton Compass on top of a field book. The item at the lower left is a GPS receiver. The item at the bottom center is a hand lens with a magnification of 10 times.

GEOBIT: GEOLOGY AND ECONOMICS

Geology encompasses far more than just the study of rocks and their history; it has economic importance as well. Other than the air we breathe everything we consume is mined or grown. Everything we consume is accessible because of the rocks that make up the local geology.

The most noticeable resource in Illinois is the agricultural use of the land. We are able to grow successful crops because of the fertile soils transported by the glaciers. Agriculture is only one resource in Illinois. Water is also very important. Most of Illinois gets it water for residential and commercial use from groundwater or Lake Michigan. Very little of the water we use is taken from rivers and reservoirs. Water is so abundant due to accessible aquifers that are prevalent throughout the State.

As much as 32 million tons of coal is mined each year in Illinois, along with about 10 million barrels of oil are removed from Central, Western, and Southern Illinois. Oil is extracted from 40 of Illinois' 102 counties. Even the rocks themselves are mined for aggregate and a lesser extent for building stone in almost every county in Northern and Southern Illinois.

Our State mineral fluorite was mined in Southern Illinois, around Hicks Dome. Fluorite deposits, also known as Fluorspar, were emplaced due to the Permian igneous activity centered on Hicks Dome. In 1942, Illinois was the leading domestic producer of fluorspar in the nation. By the 1990's cheap imported fluorite and expensive underground mining killed the domestic market. The last mine closed in December 1995. Increased fuel costs along with rising import costs in the last 20+ years are leading to renewed interest in fluorite mining in Southern Illinois.

Up until 1996, copper, zinc, and lead were mined from Northwestern Illinois. Increased land development along with decreased demand has all but stopped the metal mining industry in Illinois. Recently rare earth elements have been discovered in the igneous intrusions around Hicks Dome in concentrations that may be economically viable in the near future.

GEOBIT: ICE, WIND, AND WATER BATTLE ON LAKE MICHIGAN

February 2nd 2011 brought the worst blizzard to Northern Illinois since 1979. Over two feet of snow fell in the first couple of hours. Over 500 cars had to be abandoned on Lake Shore Drive. Steve Baumann was living in Rogers Park at the time, less than a block from the lake. On February 5th, the sun was out and if you walked to the beach you could see the power of the storm preserved in the ice. Many structures were preserved in the ice. There were mounds of frozen waves at the beach-lake margin. Frozen water spouts were common. All the features observed were formed by the power of violent winds, water, and ice during the storm.

GPS Location: 42.0129 -87.6609
A south view from 60 feet out on the frozen lake in Loyola Park showing mounds of frozen waves.

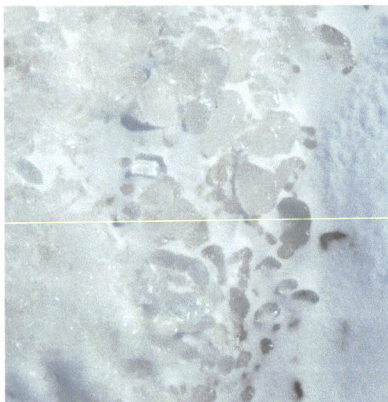

GPS Location: 42.0127 -87.6609
A close-up of ice cobbles tossed up on the frozen waves by the winds during the storm. The largest cobble is about six inches long.

GPS Location: 42.0118 -87.6607
An east view from 90 feet out on the lake, showing icebergs braking off into Lake Michigan. Notice the stream in the lower left that has carved its way through the top of the ice at the lake as the ice further west melts.

Location: 42.0113 -87.6598
During the blizzard, high winds created large water spouts. The ones near the shore became frozen, forming inverted ice cones. This one was the largest at about five feet tall, looking south.

.GEOBIT: KARST TERRAIN IS NOT OUT OF SIGHT OUT OF MIND

Karst terrain (also known as karst area or karst topography) is an area with abundant caves, sinkholes, and springs. Illinois actually has a lot of areas with karst topography. In total, karsts cover more than 10% of the state. They are mostly outside of glaciated areas of Northwestern, Western, and Southern Illinois. Karst topography includes subterranean caves, which can form beautiful structures, such as stalactites and stalagmites. Other structures can form as limestone is dissolved and precipitated elsewhere, such as nodules of limestone that form strange and very intricate structures.

Karst terrain forms when groundwater dissolves carbonate rock in the subsurface. From that caves and springs can form but so can sinkholes. A sinkhole forms when a shallow subsurface cavern collapses, usually from natural processes. Manmade sinkholes can form above abandoned underground mines and in utility corridors. Sinkholes are dangerous geologic hazards. They can form in a matter of minutes to hours and can cause great damage to people and property.

Most naturally occurring sinkholes are small (less than 100 feet wide and less than 10 feet deep) and can give an area a heavily cratered appearance. There is one sinkhole in Hardin County that tops them all. The "Big Sink" is two miles long and one mile wide. It is the largest sinkhole in Illinois and may be the largest in North America. It sits just northwest of the IL-146 and IL-1 junction with its center near GPS coordinates 37.5062° by -88.1875°. It is likely a series of smaller connected sinkholes that have formed over a single cave network.

A two and a half foot tall by four foot wide boulder of complexly structured limestone that likely formed in a cave. It is currently being used as a decorative stone along US-51 in Perry County.

A large water filled sinkhole off of the north side of IL-156 in Monroe County. Not all sinkholes are filled with water. This one is sometimes dry in the fall and winter months. The fact that the trees inside of it are standing straight up tells us that they grew after the sinkhole formed. Based on the size of the trees within the sinkhole, it probably formed sometime in the late 1970s or 1980s. It is about 280 feet long by 240 feet wide.

GEOBIT: METEOR IMPACTS IN ILLINOIS

Meteor impacts aren't found in isolated places far from where people live. There are many small ones scattered throughout the world. There are two recognized impacts in Illinois. Alas, both are deeply buried beneath surface deposits. You can't directly observe them, but you can drive over them.

The smaller of the two is the Glasford Structure or Glasford Disturbance. It is centered roughly at 40.57° by -89.77°, in Peoria County, east of the town of Glasford. It is only about two miles in diameter and is known only from deep borings. The size of the impacting object and its angle of entry is not known, but its age is. The meteor struck Illinois about 450 million years ago, just before the deposition of the Ordovician Maquoketa Group. We know this because the rocks down to the Maquoketa are not faulted. Rocks older than the Maquoketa are heavily faulted with a maximum displacement of 1,500 feet!

The second and much larger impact, sits just off the north end of O'Hare Airport, and is centered roughly at 42.04° by -87.88°, in Cook County, about 80 to 150 feet under the city of Des Plaines. The Des Plaines Disturbance is roughly 6.5 miles in diameter. Rocks are offset by more than 1,000 feet. The Disturbance was deeply eroded before the glaciers came through the area. As a result, we do not know its exact age. From deep borings and wells, we know the impact occurred after the Pennsylvanian but before the Quaternary glaciers covered the area with sediment. All we can say is it's older than two million years but younger than 299 million years. The bedrock contains near vertical faults, which indicate the meteor was around 1,000 feet in diameter and hit the Earth at nearly a 90° angle to the ground.

Shatter cones have been retrieved from both structures, essentially confirming that they are impact craters. Shatter cones are rocks that have "splintered" do to being hit by something with a lot of energy and form in a single moment.

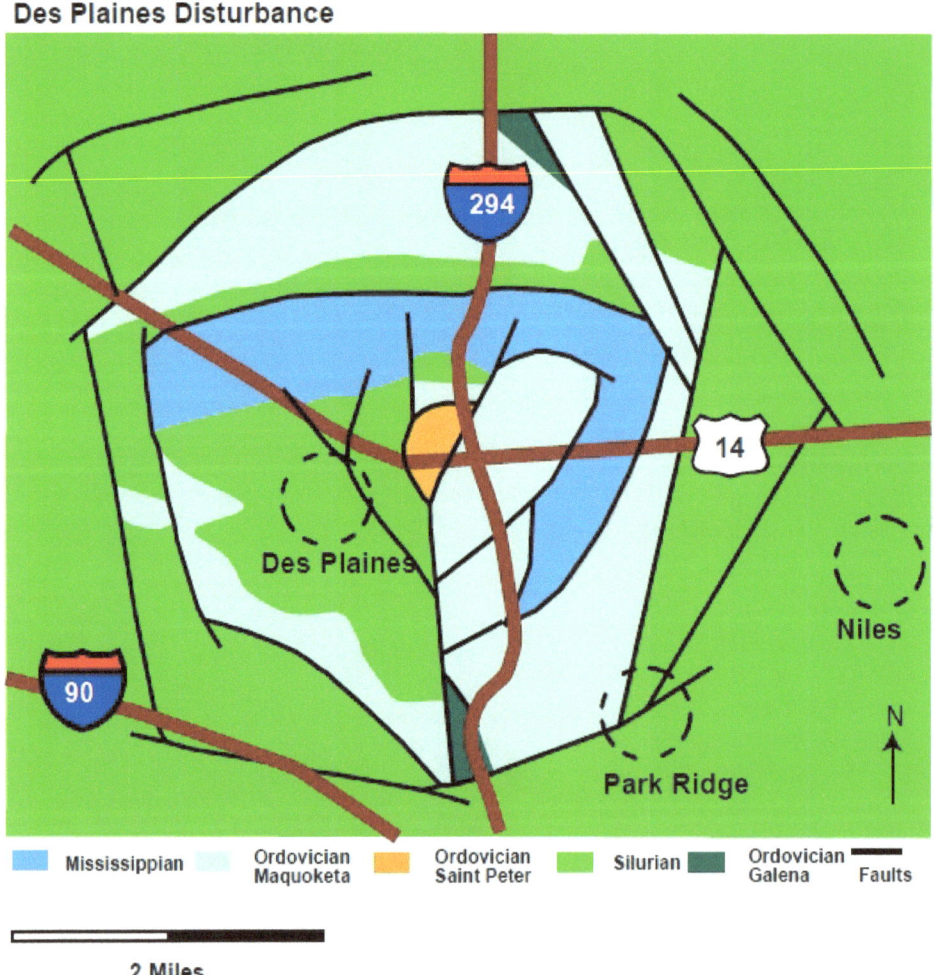

An oversimplified map of the Des Plaines Disturbance, showing the major faults and lithology of the impact crater. The Disturbance is buried under 80 to 150 feet of glacial deposits.

GEOBIT: RIVER VERSES MAN, RIVER USUALLY WINS

The flooding in Northeastern and Central Illinois in April 2013 was a record. The Illinois River reached its highest level in 70 years, cresting at around 29 feet. Although nowhere near the magnitude or scale of the spring and summer of 1993, it had many people thinking that another massive flood was on the horizon. It was a reminder of the power of flowing water and its relentless pursuit downstream. The Des Plaines River rose 15 feet in a matter of a few days in early April 2013. Major tributaries into the Mississippi River stayed at flood levels for weeks.

Flooding isn't just caused by a lot of rain. Geology and groundwater play an important role. Even though the rains stopped, some rivers continued to rise. Most rivers took a lot of initial runoff but the ground also absorbed a lot of the water. A higher groundwater table fed many streams for a couple of weeks after the flood.

In 1881 the Mississippi River actually altered its course during flooding. The town of Kaskaskia in Randolph County (the first capital city of Illinois) was left on the Missouri side of the Mississippi River after the river moved east of the town, stealing the old Kaskaskia River channel. Kaskaskia remains a part of Illinois but is only accessible by land from Missouri.

After the flood of 1993, it was decided to move the entire town of Valmeyer in Monroe County two miles northeast after flood waters destroyed the town. Most of the town was relocated between 1996 and 1998 and is now on higher ground. Ruins from the old town still stand in the floodplain.

Also in 1993, the Mississippi River broke through the Santa Fe-Fayville Levee and attempted to add Dogtooth Island to Missouri. The river cut a new channel less than two miles south of Miller City in Alexander County. This new channel path, or meander, began to form in July. It was eventually stopped at roads running perpendicular to the attempted new river path. This allowed for debris to build up and when the waters began to recede in August, the Mississippi River went back to its pre-flood course.

Kankakee River State Park in Will County. Flood waters were seven feet above normal levels. Usually you can stand on dry ground under this outcrop. Photo was taken on April 22, 2013. Compare it with the bottom photo in Volume II on page 41, which is the same outcrop when the water is at normal levels. There's about an eight foot difference!

South River Road looking north in Des Plaines, Cook County. The river is just to the right of the three foot concrete barricade in the right center of the photo. Even with the pumps going full force, it wasn't enough to hold back the water. Photo was taken on April 20, 2013.

GEOBIT: SAND ISN'T ALL THE SAME

Sandstone is a rock that is comprised of sand that is sedimentary in origin. What is sand? Sand is simply rock that has been broken down by mechanical weathering. Basically, large rocks that have been physically broken down into sand sized particles (between 0.062 and 2.000 millimeters in diameter). The natural processes involved in breaking down rock into sand can be by ice, wind, water, or gravity.

There are two main types of sands (which are subdivided further based on composition). There are arenites which contain greater than 85% sand, and wackes which are sands with 15% to 50% fines. Anything that has less than 50% sand sized particles is not considered a sand.

There is a common misconception that all sands are made of quartz. This couldn't be further from the truth. Many sands like the Saint Peter Sandstone are almost all quartz, and are comprised entirely of sand sized particles. There are also sands composed of feldspar (arkosic) or lithic grains. In Illinois, quartz and lithic sands are the most common. Lithic sands are composed of broken down rock that is not quartz or feldspar. In Illinois, they are extremely common in glacial deposits. Lithic sand composed of limestone and mudstone fragments can easily be chemically breakdown over time. This can cause undermining in foundations.

Different types of sand taken from sandstones. From upper left to upper right: gray micaceous quartz wacke (Pennsylvanian Vermilionville Sandstone in Illinois), white quartz arenite (Ordovician Saint Peter Sandstone in Illinois), yellow gray quartz arenite (Pennsylvanian in Illinois), and light brown quartz wacke (Cretaceous McNairy Formation in Illinois). From lower right to lower left: green glauconitic sublithic wacke (Cambrian Lone Rock Formation in Wisconsin), orange quartz arenite (Cambrian Galesville Sandstone in Wisconsin), purple micaceous subarkosic arenite (Precambrian Jacobsville Group in Michigan), and multicolored lithic arenite (recent beach deposits along Lake Superior at in Ontario).

Glossary

algae/algal– An organism or organisms comprised of marine dwelling single celled to complex, rootless, photosynthetic life/or having properties of algae.

alluvium– A deposit of sediments left by streams and rivers.

anhydrite – An evaporate mineral that occurs in layered deposits where large volumes of seawater have been evaporated, usually in a body of saltwater with no outlet to the sea.

anticline – A fold with layers of rock or soil sloping downward on both sides from a common ridge. Older rocks are at the center.

arenite – A sandstone or sand that contains particles of grains between 0.625 mm and 2 mm and contains less than 15% finer particles.

basement – Used to describe rock units that are below a sedimentary platform or cover and are normally ether igneous or metamorphic.

basin – A large depression that serves as the site for the accumulation of a large thickness of sediments. Beds dip towards the center from all directions. Younger rocks are towards the center.

bed/bedding – The characteristic structure of sedimentary rocks in which layers of different rock units are stacked one on top of another, in a layered sequence with oldest at the bottom and youngest at the top (in undeformed rock units).

bedrock – Rock units present beneath any surface soil, sediment or other surface cover. In some locations it may be exposed at earth's surface. In Illinois all pre-Quaternary geologic units are considered bedrock.

bituminous coal – A weak coal that contains a tar like substance. In hardness and quality, it is mid-grade.

breccia – Rock that is composed of large (over two millimeter diameter) angular fragments. The spaces between the large fragments can be filled with a matrix of smaller particles or mineral cement, binding the rock together.

calcite – An extremely common carbonate mineral and is a principle constituent of limestone and marble.

Cambrian – The first geological period/system of the Paleozoic Era of the Phanerozoic Eon.

carbonate rock – The group of rocks that contain limestone and dolostone.

Carboniferous – The fifth geological period/system of the Paleozoic Era of the Phanerozoic Eon. In North America it is split into the Mississippian and Pennsylvanian Periods/Systems.

cave/cavern – An underground void, formed naturally by a combination of processes ranging from chemical, erosion, tectonic, microorganism, pressure and atmospheric influences.

chert - A microcrystalline or cryptocrystalline sedimentary rock material composed of SiO_2, occurring as nodules, beds, and concretionary masses.

clay – A family of microscopic minerals formed from the alteration of other minerals, individual grains are less than 0.004 millimeters in diameter.

coarse grained – A descriptive word used to define rocks with particles sizes large enough to see with the naked eye.

coarse grained sand – Particles of rock between 0.5 and 1.0 millimeters in diameter.

coastal plain – An area of gently dipping rock units that were deposited near the ocean margin.

conglomerate – A sedimentary rock comprised mostly of rounded particles larger than 2 millimeters in diameter.

continental shelf – The gentle sloping margin of a continent that lies under the ocean.

convergent boundary – Plate boundary where old crust is being recycled into the mantle. They are often reflected on the surface as ocean trenches.

core – Located at the Earth's center and is divided into an inner core (solid metal), and an outer core (liquid metal). Both are composed mostly of iron and nickel. The Earth's core is about 16% of its total mass.

craton – A geologically inactive and relatively stable part of the continental crust.

Cretaceous – The third geological period/system of the Mesozoic Era of the Phanerozoic Eon.

cross beds – Beds of rock that were deposited at an angle other than horizontal. They can be flat or bowed.

crust – The solid and ridged outer shell of the Earth upon which all life lives. The two main types are the ocean crust and the continental crust. The crust accounts for slightly less than 1% of the Earth's total mass.

crypto-volcanic – A geologic structure, usually a dome, formed by subsurface igneous activity that only limitedly reaches the surface.

crystalline – A term used to describe rocks with visible crystal particles. Mostly used for carbonate and granitic rocks.

cyclothem – Alternating stratigraphic layers of marine and non-marine sediments with a basal sandstone, middle coal, and a top shale. Cyclothems are similar to a sequence, except they represent smaller units of time.

deformation – The process by which rocks undergo physical alteration through time. Often converting horizontal deposits into anticlines, synclines, domes, or basins. The process is driven by tectonics.

delta - A deposit of sediment, usually silt and clay, that forms where a stream enters a standing body of water such as a lake or ocean. When viewed from above they often appear fan or "D" shaped.

deposit – The direct settling of particles from suspension of a medium (air, water, ice, or from Earth's interior) or from direct chemical precipitation from a medium (usually water).

Devonian – The fourth geological period/system of the Paleozoic Era of the Phanerozoic Eon.

diamicton – Sediment that consists of a wide range of non-sorted to poorly sorted sediments such as sand or larger particle sizes that are suspended in a mud matrix that does not suggest a specific environment in which it formed. It is usually interchangeable with the word till.

dike – Describes a tabular or sheet like intrusion that cuts vertically or near vertical, through, and across beds of existing rock. Typically igneous in origin, but can be sedimentary in origin.

dip - The angle that a rock unit, relative to horizontal, of a planar feature, such as fault or bedding plane. The angle is measured downslope and perpendicular to the strike.

divergent boundary – Plate boundary where new crust is being created and plates are moving apart. They are often reflected on the surface as mid-ocean ridges and rift zones.

dolomite – A carbonate mineral composed of calcium magnesium carbonate, closely related to calcite.

dolostone – A sedimentary carbonate rock comprised predominately of the mineral dolomite. Often contains the same physical characteristics of limestone.

dome - An uplift that is round or elliptical in map view with beds dipping away in all directions from a central point. Older rocks are towards the center.

epicenter – The point at the Earth's surface directly above the focus of an earthquake.

erosion - A general term applied to the wearing away and movement of earth materials by gravity, wind, water, ice, chemical, or biological mechanisms.

facies - The lithological variation within a rock unit due to a change in depositional environment. Also an informal unit in stratigraphy.

fault –A break or fracture in rock along which measurable movement has occurred.

feldspar – A family of silicate minerals that originate in igneous rocks, also a major component in non-arenite sandstones.

fine grained – A term used to describe particles of rocks that are not visible with the naked eye.

fine grained sand– Particles of rock between 0.125 and 0.250 millimeters in diameter.

focus – The point within the Earth where an earthquake actually occurs.

fold – Describes layers of rock which have been bended or buckled from their original depositional position that are not faulted.

formation - A laterally continuous, somewhat homogenous rock unit with a distinctive set of characteristics that make it possible to recognize and map, from one outcrop or well to another. The basic formal rock unit in stratigraphy.

fracture – A local break along a plane in a geologic unit in which no measurable movement has occurred.

galena (mineral)– A gray metallic, heavy mineral formed predominately of lead and sulfide.

geology – The study of earth's physical structure, substance, history and the processes that act on it.

geosol – A body of sediment or rock composed of one or more soil horizons, usually applies to Quaternary soils.

glauconite – A soft mineral in the mica group. It colors sedimentary rocks green to dark green. It forms only in marine environments. Although it can occur in carbonate rock and shale, it is often associated with sandstones. The term green sand is often applied to sandstone rich in glauconite.

gneiss – A coarse grained, metamorphic rock that has a striped appearance caused by light-colored bands of granular minerals alternating with darker bands of platy or flaky minerals.

granite – A gray or pink, coarse-grained igneous rock that is made up mostly of plagioclase, potassium feldspar and quartz; but which may also contain mica or hornblende.

groundwater - Water that exists below the water table in an aquifer. Groundwater moves slowly in the between grains and fractures beneath the surface. It can also flow in underground streams.

growth fault – A type of normal fault that forms during sedimentation and typically has thicker strata on the downthrown (hanging) wall than the upthrown (foot) wall.

hardground – A term used to describe mineralization on a bedding surface that developed during a time of non-deposition.

hot spot – A large upwelling of magma from deep in the mantle, not along a plate boundary.

hydrothermal - Pertaining to the movement of hot water, the actions of hot water, or the deposits produced by the actions of hot water.

ice age – A period in Earth's history where continental sized glaciers have significantly covered the land.

igneous rock - A rock formed by the crystallization of magma or lava or by the ash that was deposited during one or more eruptions.

Illinois Episode (Illinoian) – A subdivision of the Quaternary during when sediments comprising the Illinoian glacial cycle in North America.

inland sea (continental seaway) – A shallow sea that covers central areas of continents during periods of high sea level that result in marine transgressions, also called epeiric seas.

interlobe – An area between separate, yet touching glaciers.

joint - A non-planar fracture in rock along which there has been no displacement.

Jurassic – The second geological period/system of the Mesozoic Era of the Phanerozoic Eon.

kame – A roughly round shaped hill of sand and gravel deposited at the front of a glacier as a debris fan or under a glacier.

Kansan – The youngest of the largely abandoned term used to describe pre-Illinois glacial episodes in North America.

kettle lake – A lake formed when an isolated segment of a glacier melts in the subsurface, forming a depression that becomes filled with water.

knob - A small rounded or elliptical hilltop that stands out from the surrounding hills.

limestone - A sedimentary rock consisting of at least 50% calcium carbonate ($CaCO_2$) by weight.

lithification – The physical process of turning sediment into rock.

lithology – The physical characteristics of a geologic unit such as structure, texture, composition, grain size, and color.

lithosphere - The outer layer of the Earth, that behaves rigidly. It consists of the ocean crust, the continental crust, and the upper part of the mantle. At its thickest, it is about 100 miles thick.

lobe – A tongue like projection from a continental glacier's main mass.

loess – Windblown deposits, usually of silt sized particles.

macro life – Multicellular life larger than microscopic in size.

magma – Molten rock material that occurs below Earth's surface.

mantle – A major subdivision of Earth's internal structure that exists in a semi-solid state. Located between the bottom of the crust and overlying the core. It constitutes about 83% of the planet's mass. Except for the lithosphere, it is plastic flowing rock. Also a term used to describe a thin layer of material over another type of material.

marble – A non-foliated metamorphic rock that is produced from the metamorphism of carbonate sedimentary rocks.

marine – Relating to the ocean and the activities that occur within.

meander – A natural bend in a stream or river that is "U" shaped.

member – A formal stratigraphic term used to classify small groups of similar rock in a localized area. It stratigraphy it is one rank lower than the formation.

Mesozoic – An era of geologic history that encompasses the Triassic, Jurassic, and Cretaceous Periods/Systems.

mica – A group of chemically and physically related minerals containing aluminum silicate, common in all three rock types.

micaceous – A term used to describe a rock containing significant amounts of one or more minerals of the mica group. In sandstones it tends to occur in small silvery flakes that look like glitter.

mineral – A solid, natural occurring substance, with a preferred molecular structure. They often form crystals.

Mississippian – The lower geologic subdivision of the Carboniferous Period/System.

mudstone – A sedimentary rock composed of clay and silt-size particles usually lacking defined bedding or lamination. Also used to describe sandy clays.

Nebraskan – The oldest of the two largely abandoned terms used to describe pre-Illinois glacial episodes in North America.

normal fault – A fault with net vertical movement that has occurred along a break that has formed a planar surface. The block above the fault has moved down relative to the block below the fault.

Ordovician – The second geological period/system of the Paleozoic Era of the Phanerozoic Eon.

oolite – A small sphere of carbonate or iron minerals, no more than a few millimeters in diameter and with a concentric internal structure . They primarily form by inorganic precipitation of calcium carbonate in very thin layers around a grain of sand or a particle of shell or coral.

outcrop – An exposure of a geologic unit(s) that is relatively free of vegetation. Outcrops can be formed naturally (streams and moving water) or by human action (road, rail, and building cuts).

oxbow – A short lived, crescent-shaped lake that forms when a meandering stream abandons a meander for a new course. Course changes frequently occur during flood events when overbank waters erode a new channel.

paleosol – An ancient soil that reflects exposure to the surface. Most do not exhibit direct organic activity but are largely chemically weathered surfaces. Used to describe pre-Quaternary soils.

Paleozoic – An era of geologic history that encompasses the Cambrian, Ordovician, Silurian, Devonian, Carboniferous, and Permian Periods/Systems.

Pennsylvanian – The upper geologic subdivision of the Carboniferous Period/System.

Period – A formal division of geologic time.

Permian – The sixth geological period/system of the Paleozoic Era of the Phanerozoic Eon.

Phanerozoic – A geologic eon beginning at the start of the Cambrian and continuing through today.

Pleistocene – The first epoch of the Quaternary Period, in which the all of the recent ice ages occurred.

plate – A physical ridged area of the Earth's crust, bordered by spreading, collision, and subduction zones. They can be composed of ocean crust, continental crust, or a combination of both.

pluton – An igneous rock formed deep beneath the surface of the Earth by consolidation of magma.

porphyritic – A texture term used to describe rocks containing distinct embedded crystals or crystalline particles.

porous – A term used to describe rocks containing numerous visible void space or pores.

Precambrian – A division of geologic time that encompasses all of Earth's history from its formation up to the beginning of the Phanerozoic Eon, encompassing ~88% of the Earth's history.

Pre-Illinois Episode (Pre-Illinoian) – Used by geologists to refer to the glacial advances and retreats before the Illinois Episode. It includes the old Nebraskan and Kansan glacial cycles.

quartz – A mineral composed entirely of silica (SiO_2), it is the most common rock-forming minerals at the Earth's surface.

quartzite – Usually a crystalline metamorphic rock formed by the alteration of quartz rich sandstone by heat and pressure, also can be of igneous or contact metamorphic in origin.

Quaternary – The most recent geological period/system of the Cenozoic Era of the Phanerozoic Eon.

reverse fault – A fault with vertical movement and an inclined fault plane. The block above the fault has moved upwards relative to the block below the fault.

road cut – A manmade cut in Earth's surface for the purpose of building a road.

rock – A hard aggregate of one or more minerals.

sand – A sedimentary rock composed of sand-sized particles (0.062 to 2 millimeters in diameter). Usually comprised of quartz but can contain feldspar, lithic fragments, and impurities such as silt and clay.

sandstone – A sedimentary rock composed of sand.

sediment – A loose, unconsolidated deposit of natural weathered debris, chemical precipitates, or biological debris that accumulates on Earth's surface.

sedimentary rock – A rock formed from the consolidation of sediment, usually in layered deposits or debris flows.

sequence – A series of rocks deposited in a marine transgressive to regressive environment. They are separated by unconformities. The term usually applies to sedimentary rocks but has also been used for igneous rocks.

shale – A clastic sedimentary rock that is made up of clay-size (less than 0.004 millimeter in diameter) particles. It typically breaks into thin flat pieces and usually laminated.

silt – A clastic sedimentary rock that forms from silt-size (between 0.004 and 0.062 millimeter diameter) particles.

Silurian – The third geological period/system of the Paleozoic Era of the Phanerozoic Eon.

Sloss sequence – Also known as a cratonic sequence, was proposed by Lawrence Sloss in 1963 to name transgressional seas that covered land on a continental scale.

sorting – The distribution of grain sizes within a rock. The more grains that are all the same size, the better sorted the rock is. A rock with similar grain size throughout is well sorted. A rock with a large variety of grain sizes is poorly sorted.

stream – The term is a matter of perspective but is usually used to refer to a small, narrow river. Also, a generic term for ancient rivers.

strike – A straight line that connects two points of the same elevation on a planar surface. It is always perpendicular to dip.

strike-slip fault – A fault on which net horizontal (instead of vertical) displacement occurs, typically caused by shear stress.

stromatolite – A carbonate or silica mat or mound-shaped fossil that forms from the repetitious layering of algal mat covered by trapped sediment particles. The oldest form of life recognizable in the field. They still exist today.

supercontinent – A large landmass that forms from the convergence of multiple continents, caused by Plate Tectonics.

syncline – An open, trough-shaped fold with youngest strata in the center.

System – A set of rocks deposited within a defined timeframe, equivalent to a Period.

tectonic(s) – The processes that move and deform Earth's crust.

Tertiary – The first geological period/system of the Cenozoic Era of the Phanerozoic Eon. The term was abandoned and what was the Tertiary has been divided into the Paleogene (23 to 66 million years ago) and the Neogene (2.6 to 23 million years ago).

tongue – A wedge shaped geologic unit in between another unit. A formal division of stratigraphy below the rank of formation.

thrust fault - A type of reverse fault that has a dip of less than 45°.

Triassic – The first geological period/system of the Mesozoic Era of the Phanerozoic Eon.

tufa – A variety of limestone formed by the precipitation of carbonate minerals from ambient temperature bodies of water, usually from groundwater and springs. It is also known as travertine.

till – An unsorted sediment deposited directly by a glacier and not reworked by melt water. Till is a form of diamicton.

unconformity – A contact between two rock units of different ages in which there is a span of time missing from the rock record.

vesicle/vesicular – A small and usually spherical cavity in a rock or mineral, formed by expansion of a gas or vapor before the enclosing body solidified. Also the tiny spaces in between fossils of a carbonate rock.

vug/vuggy - A small cavity in a rock. It is a larger version of a vesicle. They are often contain geodes.

weathering – A chemical or mechanical process, which breaks down rocks to smaller pieces and eventually deposited as sediment.

Wisconsin Episode – A subdivision of the Quaternary during when sediments comprising the Wisconsin glacial cycle in North America. It is the most recent of all glacial cycles.

Additional Reading

There is a great multitude of information on the geology of Illinois. The Illinois State Geological Survey now has all of its geologic maps and many of publications on its website to download and view for free. The Illinois State Museum has publications on fossils and the paleo-cultures of Illinois. The Midwest Institute of Geosciences and Engineering does not strictly focus on Illinois, but has many publications and maps pertaining to the Prairie State. Listed below are just a few of the sources of great information on the geology of Illinois.

Books:

Geology of Illinois (2010)

Geology Underfoot in Illinois (1997)

Guide to the Illinois Caverns State Natural Area (2004)

Richardson's Guide to the Fossil Fauna of Mazon Creek. (1997)

Websites:

Illinois State Geological Survey: www.isgs.illinois.edu

Illinois State Museum: www.museum.state.il.us

Midwest Institute of Geosciences and Engineering: www.mige-web.org

References

Atherton, E. Department of Registration and Education, Division of the State Geological Survey. (1947). *Some chester outcrop and subsurface sections in southeastern illinois* (Circular No. 144). Urbana, Illinois. 122-131.

Barrows, H. H. (1910). *Geography of the middle Illinois valley*. (Vol. 15). Urbana, Illinois: State Geological Survey.

Baxter, J. Department of Registration and Education, Division of the State Geological Survey. (1960). *Salem limestone in southwestern illinois* (Circular No. 284). Urbana, Illinois:

Bell, A. H. Department of Registration and Education, Division of the State Geological Survey. (1941). *Role of fundamental geologic principles in the opening of the illinois basin* (Circular No. 75). Urbana, Illinois:

Bell, A. H. Department of Registration and Education, Division of the State Geological Survey. (1961). *Underground storage of natural gas in illinois* (Circular No. 318). Urbana, Illinois:

Berg, R. C., Kempton, J. P., Follmer, L. R., & McKenna, D. P. Illinois Department of Energy and Natural Resources, Illinois State Geological Survey. (1985). *Illinoian and wisconsinan stratigraphy and environments in northern illinois: the altonian revised* (Guidebook Series No. 19). Urbana, Illinois:

Berg, R. C., McKay III, E. D., Goble, R. J., & Wang, H. Illinois Department of Natural Resources, Illinois State Geological Survey. (2013). *age of the winnebago formation of north-central illinois as determined by optically stimulated luminescence dating* (circular no. 580). Champaign, IL:

Bretz, J. H. Department of Registration and Education, Division of the State Geological Survey. (1939). *Geology of the chicago region* (Bulletin No. 65, Part I). Urbana, Illinois:

Bretz, J. H. Department of Registration and Education, Division of the State Geological Survey. (1955). *Geology of the chicago region* (Bulletin No. 65, Part II). Urbana, Illinois:

Bretz, J. H., & Harris, Jr., S. E. Department of Registration and Education, Illinois State Geological Survey. (1961). *Caves of illinois* (Report of Investigations 215). Urbana, Illinois:

Bristol, H. M., & Howard, R. H. Department of Registration and Education, Illinois State Geological Survey. (1971). *paleogeologic map of the sub-pennsylvanian chesterian (upper mississippian) surface in the illinois basin* (Circular No. 458). Urbana, Illinois:

Chrzastowski, M. J. Illinois State Geological Society, (2009). *The chicago river - a legacy of glacial and coastal processes* (Guidebook Series No. 37). Urbana, Illinois:

Cluff, R. M., Reinbold, M. L., & Lineback, J. A. Department of Registration and Education, Illinois State Geological Survey. (1981). *The new albany shale group of illinois* (Circular No. 518). Urbana, Illinois:

Curry, B. B., Graese, A. M., Vaiden, R. C., Bauer, R. A., Schumaher, D. A., Norton, K. A., Dixon, Jr., W. G., & Reed, P. C. Illinois Department of Energy and Natural Resources, Illinois State Geological Society. (1988). *Geological-geotechnical studies for siting the superconducting super collider in illinois: results of the 1986 test drilling program* (Environmental Geology Notes No. 122). Urbana, Illinois:

Dey, W. S., Davis, A. M., Curry, B. B., Keefer, D. A., & Abert, C. C. Illinois State Geological Society, (2007). *Kane county water resources investigatoins; final report on geologic investigations* (ISGS Open File Series 2007-7). Champaign, IL:

Ekblaw, G. E. Department of Registration and Education, Division of the State Geological Survey. (1938). *Kankakee arch in illinois* (Circular No. 40). Urbana, Illinois.

Emrich, G. H. Department of Registration and Education, Illinois State Geological Survey. (1966). *Ironton and galesville (cambrian) sandstones in illinois and adjacent areas* (Circular No. 403). Urbana, Illinois:

Frankie, W. T., Kolata, D. R., & Berg, R. C. Illinois Department of Energy and Natural Resources, Illinois State Geological Survey. (1999). *Guide to the geology of the rock cut state park and rockford area, winnebago county, illinois* (Field Trip Guidebook 1999C, 1999D). Urbana, Illinois:

Frankie, W. T., & Nelson, R. S. Illinois Department of Natural Resources, Illinois State Geological Survey. (2002). *Guide to the geology of the apple river canyon state park and surrounding area of northeastern jo daviess county, illinois*. Urbana, Illinois:

Frye, J. C., Willman, H. B., & Glass, H. D. Department of Registration and Education, Illinois State Geological Survey. (1964). *Cretaceous deposits and the illinoian glacial boundary in western illinois* (Circular No. 364). Urbana, Illinois:

Graeses, A. M. Department of Registration and Education, Illinois State Geological Survey. (1991). *facies analysis of the ordovician maquoketa group and adjacent strata in kane county, northeastern illinois* (Circular No. 547). Urbana, Illinois:

Grimley, D. A., & Phillips, A. C. Illinois State Geological Society, (2011). r*idges, mounds and valleys: glacial-interglacial history of the kaskaskia basin, southwestern illinois* (ISGS Open File Series 2011-1). Champaign, IL:

Hansel, A. K., Berg, R. C., Philips, A. C., & Gutowski, V. G. Illinois Department of Energy and Natural Resources, Illinois State Geological Survey. (1999). *glacial sediments, landforms, paleosols, and a 20,000 year old forest bed in east-central illinois* (Guidebook Series No. 26). Urbana, Illinois:

Hansel, A. K., & Johnson, W. H. Department of Registration and Education, Illinois State Geological Survey. (1996). Wedron and mason groups: Lithostratigraphic reclassification of deposits of the wisconsin episode, lake michigan lobe area (Bulletin No. 104). Urbana, Illinois:

Hensel, B. Illinois Department of Energy and Natural Resources, Illinois State Geological Survey. (1992). *Natural recharge of groundwater in illinois* (Environmental Geology Notes No. 143). Urbana, Illinois:

Hughes, G. M., Kraatz, P., & Landon, R. A. Department of Registration and Education, Illinois State Geological Survey. (1966). *Bedrock aquifers of northeastern illinois* (Circular No. 406). Urbana, Illinois:

Johnson, W. H., Follmer, L., Gross, D. L., & Jacobs, A. M. Illinois State Geological Survey, (1972). *Pleistocene stratigraphy of east-central illinois* (Guidebook Series No. 9). Urbana, Illinois:

Johnson, W. H., Hansel, A. K., Bettis III, E. A., Karrow, P. F., Larson, G. J., Lowell, T. V., & Schneider, A. F. Illinois Department of Natural Resources, Illinois State Geological Survey. (1996). *late quaternary temporal and event classifications, great lakes region, north america* (Reprint series 1997B). Champaign, IL:

Kimball Brown, M. (1975). The Zimmerman site: Further excavations at the grand village of kaskaskia. *Illinois State Museum: Reports of Investigation*, (32),

King, J. E. (1982). Fossils. *Illinois State Museum: Story of Illinois Series*, (14),

Kolata, D. R. Illinois State Geological Survey, (1991). *Tippecanoe i subsequence; middle and upper ordovician series* (Reprint series 1991-T5). Champaign, IL:

Kolata, D. R. Illinois State Geological Survey, (1991). *Tippecanoe sequence overview; middle ordovician series through lower devonian series* (Reprint series 1991-T4). Champaign, IL:

Kolata, D. R., & Buschbach, T. C. Department of Registration and Education, Illinois State Geological Survey. (1976). *plum river fault zone of northwestern illinois* (Circular No. 491). Urbana, Illinois:

Kolata, D. R., Buschbach, T. C., & Treworgy, J. D. Department of Registration and Education, Illinois State Geological Survey. (1978). *The sandwich fault zone of northern illinois* (Circular No. 505). Urbana, Illinois:

Kolata, D. R., & Nimz, C. K. (2010). *Geology of Illinois*. Champaign, IL: University of Illinois Board of Trustees.

Kolata, D. R., Treworgy, J. D., & Masters, J. M. Department of Registration and Education, Illinois State Geological Survey. (1981). *Structural framework of the Mississippi embayment of southern Illinois.* (Circular No. 516). Urbana, Illinois:

Leetaru, H. E., Sargent, M. L., & Kolata, D. R. Illinois Department of Natural Resources, Illinois State Geological Survey. (2004). *Geologic atlas of cook county for planning purposes*. Champaign, IL:

Mast, R. F., Ruch, R. R., & Meents, W. F. Department of Registration and Education, Illinois State Geological Survey. (1973). *vanadium in devonian, silurian, and ordovician crude oils of illinois* (Circular No. 483). Urbana, Illinois:

McGinnis, L. D. Department of Registration and Education, Illinois State Geological Survey. (1966). *Crustal tectonics and precambrian basement in northeastern illinois* (Report of Investigations 219). Urbana, Illinois:

Mehnert, E. Illinois Department of Natural Resources, Illinois State Geological Survey. (2010). *Groundwater flow modeling as a tool to understand watershed geology: blackberry creek watershed, kane and kendall counties, illinois* (circular no. 576). Champaign, IL:

Mehnert, E., Hackley, K. C., Larson, T. H., Panno, S. V., Pugin, A., Wehmann, H. A., Holm, T. R., & Roadcap, G. S. Illinois State Geological Society, Illinois State Water Survey. (2004). *The mahomet aquifer: recent advances in our knowledge* (ISGS Open File Series 2004-16). Champaign, IL

Melhorn, W. N. Illinois Department of Energy and Natural Resources, Illinois State Geological Survey. (1991). *Tippecanoe i subsequence; middle and upper ordovician series* (Reprint series 1991L). Champaign, IL:

Mikulic, D. G., & Kluessendori, J. Illinois Department of Energy and Natural Resources, Illinois State Geological Survey. (1999). *the classic silurian reefs of the chicago area* (Guidebook Series No. 29). Urbana, Illinois:

Nelson, W. J. Department of Natural Resources, Illinois State Geological Survey. (2002). *Sequence stratigraphy of the lower chesterian (mississippian) strata of the illinois basin* (Bulletin No. 107). Champaign, Illinois:

Nelson, W. J. Department of Registration and Education, Illinois State Geological Survey. (1995). *Structural features in illinois* (Bulletin No. 100). Urbana, Illinois:

Nelson, W. J., & Lumm, D. K. Department of Registration and Education, Illinois State Geological Survey. (1987). *Structural geology of southeastern Illinois and vicinity.* (Circular No. 538). Urbana, Illinois:

Nelson, W. J., & Welbel, C. P. Department of Registration and Education, Illinois State Geological Survey. (1996). *Geology of the lick creek quadrangle johnson, union, and williamson counties, southern illinois* (Bulletin No. 103). Urbana, Illinois.

Potter, P. E. Department of Registration and Education, Division of the State Geological Survey. (1962). *Late mississippian sandstones of illinois* (Circular No. 340). Urbana, Illinois:

Potter, P. E. Illinois State Geological Survey, (1963). *Late paleozoic sandstones of the illinois basin* (Report of Investigations 217). Urbana, Illinois:

Reed, P. C. Illinois State Geological Society, (1974). *Data from controlled drilling program in boone and dekalb counties, illinois* (Environmental GEology Notes No. 77). Urbana, Illinois:

Reinertsen, D. L. Illinois Department of Energy and Natural Resources, Illinois State Geological Survey. (1992). *Guide to the geology of the galena area* (Field Trip Guidebook 1992B). Urbana, Illinois:

Reinertsen, D. L., Jacobson, R. J., Killey, M. M., Nelson, R. S., & Reed, P. C. R. (1993). *Guide to the geology of the Lewistown-spoon river area Fulton County, Illinois*. Champaign, IL: Illinois State Geological Survey.

Reynolds, R. L., Goldhaber, M. B., & Snee, L. W. (1997). Paleomagnetic and 40ar/39ar results from the Permian Downeys Bluff Sill - evidence for Permian igneous activity at Hicks Dome, southern Illinois Basin. In R. W. Scott Jr. (Ed.), *Evolution of Sedimentary Basins - Illinois Basin*. Denver, CO: U.S. Department of the Interior.

Rogers, C., (2019). *The mysterious Tully Monster just got more mysterious*. Live Science News Release

Sargent, M. L. Illinois State Geological Survey, (1991). *Sauk sequence; cambrian system through lower ordivican series*. Champaign, IL:

Shrode, R. S. Department of Registration and Education, Division of the State Geological Survey. (1948). *Unusual oolite grains from the ste. genevieve limestone* (Circular No. 144). Urbana, Illinois. 140-144.

Stohr, C. J., Petras, J., Mikulic, D. G., & Thomason, J. (2011). Stereophotographic measurement of joint and bedding orientation at thornton quarry, illinois. In Illinois State Geological Survey.

Stumpf, A. J., Hansel, A. K., & Barnhardt, M. L. (2003). *Geologic mapping of glacial drift aquifers in the greater Chicago area of Illinois*. Champaign, IL: Illinois State Geological Survey.

Swann, D. H., Lineback, J. A., & Frund, E. Department of Registration and Education, Illinois State Geological Survey. (1965). *The borden siltstone (mississippian) delta in southwestern illinois* (Circular No. 386). Urbana, Illinois:

Templeton, J. S., Graf, D. L., Horberg, L., & Workman, L. E. Department of Registration and Education, Division of the State Geological Survey. (1951). *Short papers on geologic subjects* (Circular No. 170). Urbana, Illinois.

Templeton, J. S., & Willman, H. B. Department of Registration and Education, Illinois State Geological Survey. (1963). *Champlainian series (middle ordovician) in illinois* (Bulletin No. 89). Urbana, Illinois.

Treworgy, J. D. Department of Registration and Education, Illinois State Geological Survey. (1988). *The illinois basin - a tidally and tectonically influenced ramp during mid-chesterian time* (Circular No. 544). Urbana, Illinois:

Tri-State Committee on Correlation of the Pennsylvanian System in the Illinois Basin. Illinois State Geological Survey, Indiana. Geological Survey, & Kentucky Geological Survey, (2001). *Toward a more uniform stratigraphic nomenclature for rock units (formations and groups) of the Pennsylvanian system in the Illinois basin*. Illinois Basin Consortium.

Udden, J. A. (1914). *Some deep borings in illinois*. (Vol. 24). Urbana, Illinois: Illinois State Geological Survey.

Vaiden, R. C., Smith, E. C., & Larson, T. H. Illinois Department of Natural Resources, Illinois State Geological Survey. (2004). *groundwater geology of dekalb county, illinois with emphasis on the troy bedrock valley* (Circular No. 563). Urbana, Illinois:

Visocky, A. P., Sherrill, M. G., & Cartwright, K. Illinois Department of Energy and Natural Resources, Illinois State Geological Survey, Illinois State Water Survey. (1985). *Geology, hydrology, and water quality of the cambrian and ordovician systems in northern illinois* (Cooperative Groundwater Report 10). Urbana, Illinois:

Walter, P. (1963). Transactions of the Illinois state academy of science. In *Transactions of the Illinois State Academy of Science* (Vol. 56, pp. 59-67). Urbana, Illinois: Illinois State Geological Survey.

Webb, N. D., Grimley, D. A., Phillips, A. C., & Fouke, B. W. (2012). Origin of glacial ridges (ois 6) in the Kaskaskia Sublobe, southwestern Illinois, USA. *Quaternary Research, (78), 341-352.*

Weller, J. M. Department of Registration and Education, Division of the State Geological Survey. (1943). *Rhythms in upper pennsylvanian cyclothems* (Circular No. 92). Urbana, Illinois:

Whitaker, S. T. Illinois Department of Energy and Natural Resources, Illinois State Geological Survey. (1988). *Ramp-platform model for silurian pinnacle reef distribution in the illinois basin.* Urbana, Illinois:

Whiting, L. L., & Stevenson, D. L. Department of Registration and Education, Illinois State Geological Survey. (1965). *The sangamon arch* (Circular No. 383). Urbana, Illinois.

Wiggers, R., Mountain Press Publishing Company. (1997). *geology underfoot in illinois.* Missoula Montana

Willman, H. B. Department of Registration and Education, Illinois State Geological Survey. (1973). *rock stratigraphy of the silurian system in northeastern and northwestern illinois* (Circular No. 479). Urbana, Illinois:

Willman, H. B. Department of Registration and Education, Illinois State Geological Survey. (1971). *summary of the geology of the chicago area* (Circular No. 460). Urbana, Illinois:

Willman, H. B., Atherton, E., Buschbach, T. C., Collinson, C., Frye, J. C., Hopkins, M. E., Lineback, J. A., & Simon, J. A. Department of Registration and Education, Illinois State Geological Survey. (1975). *Handbook of illinois stratigraphy* (Bulletin No. 95). Urbana, Illinois.

Willman, H. B., & Frye, J. C. Department of Registration and Education, Illinois State Geological Survey. (1969). *high-level glacial outwash in the driftless area of northwestern illinois* (Circular No. 440). Urbana, Illinois:

Willman, H. B., & Frye, J. C. Department of Registration and Education, Illinois State Geological Survey. (1970). *Pleistocene stratigraphy of illinois* (Bulletin No. 94). Urbana, Illinois:

Willman, H. B., & Kolata, D. R. Department of Registration and Education, Illinois State Geological Survey. (1978). *the platteville and galena groups in northern illinois* (Circular No. 502). Urbana, Illinois:

Willman, H. B., Lowenstam, H. A., & Workman, L. E. Illinois State Geological Survey. (1950). *Guidebook: Field conference on niagaran reefs in the chicago region.* Urbana, Illinois:

Willman, H. B., & Payne, J. N. Department of Registration and Education, Division of the State Geological Survey. (1943). *Early ordovician strata along fox river in northern illinois* (Circular 100). Urbana, Illinois:

Willman, H. B., & Payne, J. N. Department of Registration and Education, Division of the State Geological Survey. (1942). *Geology and mineral resources of the marseilles, ottawa, and streater quadrangles* (Bulletin No. 66). Urbana, Illinois:

Workman, L. E. Department of Registration and Education, Division of the State Geological Survey. (1938). *The preglacial rock river valley as a source of groundwater for rockford* (Circular No. 36). Urbana, Illinois:

Maps

Barnhardt, M.L., Institute of Natural Resource Sustainability, Illinois State Geological Survey (2009), *surficial geology of zion quadrangle, lake county, illinois, and kenosha county, wisconsin*, Statemap Zion-SG

Baumann, S.D.J., Midwest Institute of Geosciences and Engineering (2011), *surficial geologic map of castle rock state park & lowden miller state forest, ogle county, Illinois, U.S.A.*, M-092011-1A

Baumann, S.D.J., Midwest Institute of Geosciences and Engineering (2011), *surficial geologic map of keepataw forest preserve and lemont area, will, cook, and dupage counties, Illinois, U.S.A*, M-102011-1A

Baumann, S.D.J., Midwest Institute of Geosciences and Engineering (2013), *surficial geology of the romeoville quadrangle, part of cook, dupage, and will counties, Illinois*, United States, M-112013-3A

Baumann, S.D.J., *Midwest Institute of Geosciences and Engineering (2014), surficial geologic map of the Stockton quadrangle, jo daviess county, Illinois, United States*, M-102014-1B

Denny, F.B., King, B., Mulvaney-Norris, J., Malone, D., Institute of Natural Resource Sustainability, Illinois State Geological Survey (2010), *bedrock geology of kabers ridge quadrangle, hardin, gallatin, and saline counties, Illinois*, IGQ Kabers Ridge-BG

Denny, F.B., Nelson, W.J., Munson, E., Devera, J.A., Amos, D.H., Prairie Research Institute, Illinois State Geological Survey (2011), *bedrock geology of rosiclare quadrangle, hardin county, Illinois*, Statemap Rosiclare-BG

Harrison, R.W., U.S. Department of the Interior, U.S. Geological Survey (1999), *geologic map of the thebes quadrangle, Illinois and Missouri*, MAP GQ-1779

Jacobson, R.J., Weibel, C.P., Illinois Department of Natural Resources, Illinois State Geological Survey (1993), *geologic map of the makanda quadrangle, Jackson, union, and willamson counties, Illinois*, IGQ-11

McLean, M.M., Kelly, M.D., and Riggs, M.H., Illinois State Geological Survey (1997), *Topography of the bedrock surface in McLean County, Illinois*

McLean, M.M., Kelly, M.D., and Riggs, M.H., Illinois State Geological Survey (1997), *Thickness of Quaternary deposits in McLean County, Illinois*

Murphy, C.J., Malone, D.H., and Shields, W., Illinois State University, EdMap (2014), *Surficial geologic map of the Arrowsmith 7.5' quadrangle*

Nelson, W.J., Devera, J.A., Illinois Department of Natural Resources, Illinois State Geological Survey (2007), *bedrock geology of mt. pleasant quadrangle, union and Johnson counties, Illinois*, IGQ Mt. Pleasant-BG

Riggs, M.H. and Abert, C.C., Illinois State Geological Survey (1998), *Cumulative sand and gravel thickness in McLean County, Illinois*

Disclaimer

The vast majority of the photos throughout Volumes I and II were taken in 2011-2013 during field visits. At the time this second edition is going to print, one of the authors still lives in Illinois. Sometimes sites change. Roadcuts get covered or disappear due to road expansion, parks get upgraded, floods damage outcrops, people build new structures, etc. Every reasonable effort was made to make sure all the sites visited in Volume II are up to date. If changes have occurred, we try to mention them. We are not infallible. With the primary author (Steven Baumann) no longer living in Illinois, there can be things we missed.

www.ingramcontent.com/pod-product-compliance
Lightning Source LLC
Chambersburg PA
CBHW051202220526
45473CB00003B/875